Hope you enjoy!

Ann Ashley Watters

The Complete Fly Fisher

The Great Ranch
COOKBOOK

By Gwen Ashley Walters

Illustrations by Betsy Hillis

Published by Guest Ranch Link
P.O. Box 5165
Carefree, AZ 85377

www.guestranchlink.com

Edited by Olin B. Ashley

Front Cover — The Home Ranch, Colorado
Back Cover — Crystal Creek Lodge, Alaska;
Crescent H Ranch, Wyoming; Triple Creek Ranch, Montana

Library of Congress Catalog Card Number:
98-92689

ISBN 0-9663486-0-5

Dedication

To Mom and Dad, with Love

In Memoriam – Claranelle Lively Walters

Foreword

Gwen Ashley Walters has a sensitive approach to food, is a wiz at preparation and as you will see in reading and using this book, she is both very creative and thorough — a rare combination indeed. I have known Gwen since she came to our week long Cooking School in Old Town Albuquerque, New Mexico several years ago. She was very interested in preparing all the dishes we had on the menu for each day and was quite energetic. As the week unfolded, she intimated to me that she was interested in possibly making a career change from marketing executive...to the culinary field! We chatted some about the options and before I knew it, several months later—she told me she had been accepted at the Scottsdale Culinary Institute, where she went on to graduate with honors. As is clear, Gwen has a special love for guest ranches. Her enthusiasm and interest in guest ranching and the special dining they afford makes this book a "must" for anyone yearning to visit a guest ranch!

Jane Butel —
Author of 14 cookbooks,
including "Tex-Mex Cookbook,"
"Chili Madness" and "Southwestern Kitchen"

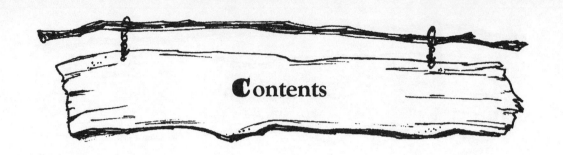

Contents

CONTENTS BY STATE AND RANCH

CONTENTS BY RECIPE CATEGORY

LAMB

FISH AND SEAFOOD

PORK AND POULTRY

VEGETABLES

STARCHES

DESSERTS

Introduction

One of the great adventures in life is visiting a guest ranch, whether the ranch is close to home or far away. There is just something "American" about saddling up for a day's ride, hiking in majestic mountains, angling grass-lined, crystal clear streams, or just cozying up in front of a roaring fire in the lodge. There are hundreds of guest ranches all over America, and their guest lists are growing year by year. Part of the "great outdoor experience" at a guest ranch is comfortable surroundings to nurse citified bodies after the day's exhausting toll. And even more rewarding is being treated to hearty food after a long day in the wilderness. Ranch cuisine ranges from ranch "grub" to familiar comfort food to elaborate platings of exotic cuisine.

The ranches presented in this book were chosen first and foremost for their outstanding cuisine. You won't find "average" food in this book. Of course, each ranch's accommodations and commitment to guest satisfaction are just as wonderful as the food. The interesting thing about all of these ranches is how much time and effort goes into their food programs. Most employ professional Chefs, many of whom have enjoyed their association for years. The ranches spend as much time planning and creating their gastronomical delights as they do any other part of their operation. Based on guest comments and the number of returning guests each year, it's not surprising.

This book brings you some of our nation's top Western Guest Ranches and Lodges, highlighting the personification of what the great escape should be. Use the book to find the destination that is perfect for your getaway, or use it to recreate the fabulous feasts you remember savoring during your last visit. If you haven't been to a guest ranch before, drop whatever you are doing, pick up the phone and call one of these ranches to book a trip. I promise, once you've visited one of these retreats, you'll be hooked for life, always reliving those wondrous memories of the great blue yonder, and already planning your next Western excursion.

Travel Thoughts to Ponder

I've done my best to accurately present the facts as well as the personal descriptions of each ranch. I hope you use this book to find a ranch or lodge that fits your idea of a dream vacation, but before you make your plans, read the following points to help you make your decision.

- All rates listed in this book represent the 1998 season.
- Call the ranch or lodge to verify the rate and ask if there are other packages you might be interested in considering.
- Ask what is included in the rate such as lodging, meals, which activities, transportation, and gratuity.
- Ask about the method of payment. Many ranches do not accept credit cards.
- If gratuity is not included in the rate, know that it is customary to leave a percentage of your total bill for the staff. The amount is up to you, based on the level of service you were provided, with 10% an "okay" tip, 15% a "good" tip and anything above 15% an excellent tip. The monies are distributed to all staff (sometimes excluding guides who are tipped directly at the time of service). A good portion of these employees' compensation comes from the service tips the ranch receives.
- If it is not clear from the description, ask the ranch about a children's program. Some ranches are more prepared for children than others. Ask what age children must be to participate in each ranch activity.
- If you smoke, ask about their smoking policy. If you drink alcoholic beverages, ask about the availability and cost of your drink of choice.
- Ask about arrival and departure times and transportation options.

These tips are only suggestions. I'm sure you will think of other questions to ask as you plan your vacation. The most important thing is to select a destination that offers the activities, food and ambiance you want. All of the ranches in this book are destinations worth visiting. Some will be more appealing to you and have more of your interests than others. My hope is that through this book, you will find vacation destinations that exceed your expectations. Let me know what your experience is, from making a reservation to spending time at a ranch, including the things you like and don't like. I want to include your input in the next edition.

Visit my Website to E-mail me: **http://www.guestranchlink.com**

I will do my best to answer any questions you may have. If I don't know the answer, I'll find someone who does.

Recipe for Success

Every single recipe in this book was tested in my home kitchen in Scottsdale. I know these recipes will work for you in your kitchen, regardless of your experience level. I adjusted the ingredient amounts when necessary, converting quantities from pounds and ounces to cups and measuring spoons in many cases. I also reduced serving quantities from 20 or 40 or 50 to a more realistic 4 or 6 or 8. I want you to have the same success and fun I had re-creating these scrumptious slices of the ranch personalities. If you run into any problems or have any questions about the recipes, ingredients or procedures, contact me through my Website (www.guestranchlink.com) and I will help you. Read through the points I've outlined below to help ensure your success. But if that doesn't help, I'm only a digital link away.

- Read the recipe all the way through before you start; even before you go to the grocery store to buy the ingredients. The directions might tell you to start a process the night before.

- Mise en Place. This is the French term for "everything in its place." It means gathering all the ingredients and equipment (and reading the entire recipe) before you begin. This will save you precious time and more than likely determine your success more than any other tip I could give you.

- Watch for the word "divided" in the ingredient list. It means that ingredient will be used more than once in the recipe. The directions will tell you how to divide the ingredient.

- Measure ingredients in the appropriate utensils. Use metal measuring cups for dry ingredients, such as flour, sugar, chopped carrots, etc. Level off with a straight edge across the top of the cup for the appropriate measurement. Use glass or plastic measuring cups with spouts for liquid ingredients such as water, milk, honey, etc. Get down to eye level for more accurate measuring.

- If the preparation method is listed before the ingredient, then you measure it prepared. For example, 1/2 cup chopped carrot means chop the carrot first, then measure 1/2 cup. If the preparation is after the ingredient, measure first and then prepare.

- All yeast called for in these recipes is active dry. 1 (.25-ounce) package is equal to 2-1/4 teaspoons.

- Set cooking and baking times for 5 to 10 minutes less than the shortest time given in a recipe.

- For goodness' sake, have fun! Experiment with ingredients, cooking times, or whatever strikes you. My favorite experiment is trying new wines with new dishes!

Measurement Conversions

3 teaspoons = 1 tablespoon
1 cup = 8 ounces (for milk, water, wine, vinegar, etc.)
1 pound = 16 ounces
2 cups = 1 pint
2 pints = 1 quart
4 quarts = 1 gallon
1 cup = 16 tablespoons

Dry ingredient weights and some "thick" liquid ingredient weights cannot be converted into cup measurements. For example 1 cup of flour is 4 ounces, not 8. And 1 cup of sugar is 6 ounces.

Common Procedures

TOASTING NUTS

Preheat oven to 350°. Spread nuts in a single layer on a baking sheet. Bake for 4-8 minutes, watching carefully so you don't burn them. It will take longer if your nuts are refrigerated or frozen. The size and fat content of the nut will impact the time it takes to toast. Look for a change in color and/or a strong nutty aroma to determine doneness.

ROASTING GARLIC (WHOLE HEAD)

Preheat oven to 350°-375°. Cut 1/3 of top (pointy end) off, exposing cloves. Rub with a small amount of olive oil. If desired, also rub with fresh chopped herbs such as rosemary, thyme and/or oregano. Place in a shallow baking pan and bake for 25 to 35 minutes, or until cloves are soft and exposed ends turn golden brown. Remove and cool slightly. Remove cloves from papery shell. Store in refrigerator in an airtight container up to 1 week.

ROASTING PEPPERS

Preheat oven to 350°-375°. Lightly oil peppers and make a tiny slit in the top or bottom to allow steam to escape. Place on a baking sheet and bake for 30 to 50 minutes, turning often. Skin will blister and look like it is separating from the flesh and may even turn brown/black. Remove and place in a large bowl. Cover with plastic wrap and let steam for 15 minutes. Peel when cool enough to handle. I prefer not rinsing in water to help remove stubborn peel. It washes away too much flavor. Instead, bake longer so peel is easier to remove. You can also blister pepper skins over a hot grill or gas flame. It takes less time but more of your attention during that time, and I don't think the roasting is as even as it is in the oven.

CLARIFYING BUTTER

There are two methods, the proper method, and mine. Mine is easier, but the result is not quite as pure. It works for me. You decide what works for you. In the proper method, you bring an amount of butter, generally 1 pound (4 sticks) to a boil over medium heat. Let it bubble until it stops crackling and popping. A foam will rise and then fall and then sort of rise again. Once the foam has fallen once, it doesn't take long for it to try to come up again. The whole process shouldn't take more than 15 to 20 minutes. And it can burn quickly, so stay close to the pan. The butter will be a golden yellow color and the foam will settle on the bottom of the pan when you turn off the heat. This foam is the milk solids. The resulting clear golden liquid is pure butterfat. Strain the butterfat through a cheesecloth-lined sieve, discarding the milk solids. Store the clarified butter in the refrigerator. It should last several weeks.

I usually don't need a whole pound, so I take a stick of butter, put it in a 2-cup glass measuring cup and put it in the microwave. I heat mine on high for 1-1/2 to 2 minutes, covered with a paper towel. You will have to adjust your time for the power of your microwave. I watch it carefully. It will bubble up and the foam will settle on the bottom. I then just pour off the golden butterfat, leaving a small amount of butterfat and the milk solids in the measuring cup. It's not as pure, as some of the solids invariably end up with the butterfat, but it works pretty well.

ALASKA

Crystal Creek Lodge

Box 92170
Anchorage, AK 99509
(800) 525-3153
Website: http://
    www.crystalcreeklodge.com
E-mail: crystal@alaska.net

Season: mid-June
through September

Capacity: 24

Accommodations: 13
comfortable rooms with private
baths all within the spacious and
sporting main lodge

Activities: Guided river and
stream fishing for salmon, trout,
char and grayling, all accessed by
float plane and helicopter

Rates: $5,200 weekly per person,
double occupancy, including
lodging, gourmet meals,
transportation to the lodge from
Dillingham, AK, and lots of
other fishing related stuff like
daily air transportation to rivers
and stream destinations, fishing
licenses, equipment to use during
your visit and a container to ship
your catch home!

Crystal Creek Lodge is a remote wilderness
adventure. While it's not a ranch in that it does
not have a horse program, it has many other
enticements that compelled me to include it in this
book. Two major reasons are Dan Michels, the
owner, and Douglas Einck, the Chef. Crystal
Creek is not just a destination. It's an experience
and one that's never forgotten by those who
venture to this remote southwestern corner of
Alaska, near Bristol Bay. And you can't forget it
because of Dan and his unrelenting dedication to
customer service and Doug, with his passion for
translating his vision of culinary artistry into
ethereal cuisine in the middle of nowhere.

Oh, and did I mention the fishing? It's among the
finest in the world, and according to some of
Dan's well-traveled guests, it is the finest. A week
at Crystal Creek will put you in remote rivers and
streams accessible only through the lodge's vintage
float planes or its Sikorsky S-55 helicopter. Once
at your destination, you may journey deeper into
the wilderness in a jet-powered riverboat designed
to navigate the shallowest waters or you may wade
out into the clear flowing water. Whether you
boat, drift, bank-walk or stream-wade is
determined not so much by what you prefer but
what fish you are pursuing.

Five species of Pacific Salmon migrate from the
ocean into the rivers and streams of this arcane
corner of the world to complete their life cycle. In
addition, 6 native freshwater species thrive in the
wake of the salmon cycle. You pick the week of
fishing you want, either based upon what is most
convenient to your schedule or, more likely, what

fish are running when. The amount of research, planning and plotting that occurs to track this phenomenon of nature is astounding. Of course, all of the effort is not just to please fish-happy city-dwellers. It's part of an overall ecosystem management program that fortifies and protects the delicate cycle of life for these migrating game fish.

The lodge's goal is to ensure your non-fishing time is just as memorable through genuine hospitality and attention to detail. From the wader-drying room to fly-tying stations to overstuffed couches in front of roaring fires, Dan and his professional staff have thought of everything to make your stay unforgettable. One of the most important goals Dan has accomplished is hiring and retaining Chef Einck for seven consecutive seasons.

Classically trained and professionally seasoned in some of the country's finest resorts and hotels, Chef Einck masterfully executes fine dining cuisine in spite of his remote location and limited access to products. One product that is never a problem is fresh salmon. Chef Einck and his small staff offer three entrees each evening, two of which are traditional classics, such as Veal Florentine or Beef Wellington while the third is "comfort food" such as Southwestern Enchiladas or Pork Chops and Applesauce. A pastry chef at heart, he prepares desserts that are as fabulous as the entrée. You might see Sacher Torte, Alsatian Apple Tart or Decadent Chocolate Encased Pear.

Chef Einck enjoys working at Crystal Creek because of the returning clients who year after year provide positive feedback on the product he creates. "It's important to me to know that they come to the lodge for more than just the fishing," he says. Chef Einck has shared some wonderful dishes with us. It's no wonder the guests shower him with praise. His food is fabulous and even more so given the challenges of his remote location.

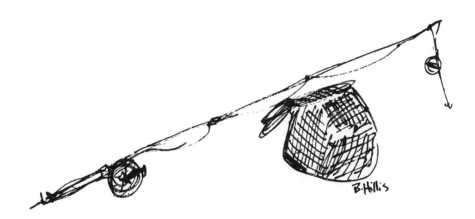

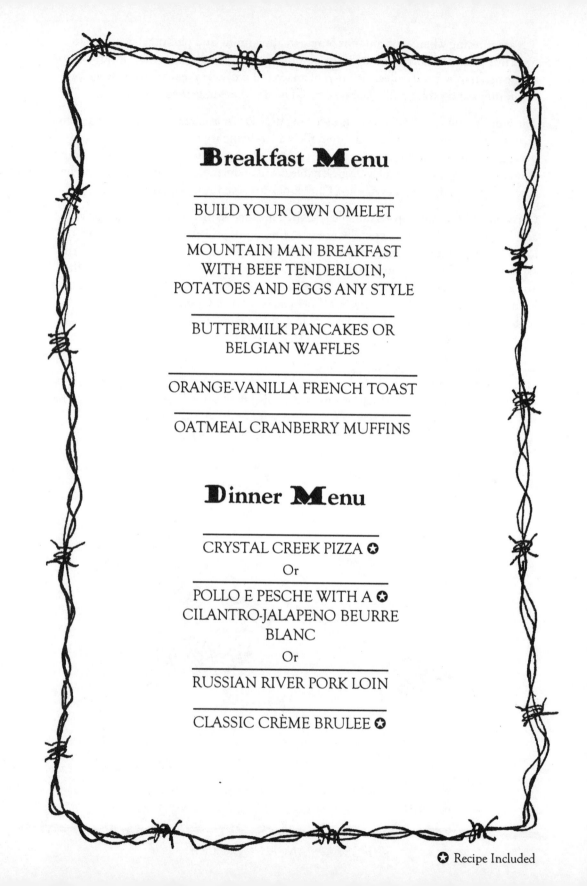

Breakfast Menu

BUILD YOUR OWN OMELET

MOUNTAIN MAN BREAKFAST
WITH BEEF TENDERLOIN,
POTATOES AND EGGS ANY STYLE

BUTTERMILK PANCAKES OR
BELGIAN WAFFLES

ORANGE-VANILLA FRENCH TOAST

OATMEAL CRANBERRY MUFFINS

Dinner Menu

CRYSTAL CREEK PIZZA ✪

Or

POLLO E PESCHE WITH A ✪
CILANTRO-JALAPENO BEURRE
BLANC

Or

RUSSIAN RIVER PORK LOIN

CLASSIC CRÈME BRULEE ✪

✪ Recipe Included

Crystal Creek Pizza

I could easily eat the crust and be totally happy . With the toppings, I'm
ecstatic. The crust is fun to make and the pizza sauce couldn't be easier.
Chef Einck's selection of toppings is kind of avant-garde, though you could
top this any way you like. His choices are colorful and flavorful. It's the
kind of pizza that deserves the best micro-brew you can find, or a nice,
chilled Pinot Grigio.

YIELD = 5 (8") PIZZAS

Roasted Garlic and Herb Dough:
3 cups bread flour
3/4 to 1 cup water
1 package quick-acting yeast (.25-ounce)
1 tablespoon sugar
1-1/2 teaspoons dried basil

1 teaspoon black pepper
1 teaspoon dried thyme
1/4 teaspoon dried oregano
1 tablespoon roasted garlic*
1 tablespoon + 2 teaspoons olive oil
3/4 teaspoon salt

Combine all ingredients in a large mixing bowl. With the dough hook, mix on low until
ingredients are incorporated, about 2 minutes, then switch to medium speed and mix for
6 to 8 minutes. Turn dough into a greased bowl and cover. Let rise in a warm spot until
doubled in size, about 1 to 2 hours. Turn dough onto a lightly floured surface and punch
down to remove all air. Portion into 5 equal size balls. Cover with plastic wrap.

Preheat oven to 350°. One at a time, keeping the others covered, roll the dough into a 9"-
10" circle. (Chef Einck actually spins the dough to the size of a dinner plate.) Brush with
olive oil and sprinkle with a little salt. Bake until the dough just starts to turn golden
brown, about 10 to 12 minutes. Bake as many as you plan on using. Refrigerate unused
dough for up to 2 days or freeze for up to 1 month. The pre-baked crust is now ready for
the sauce and toppings.

*To roast garlic, see page 11.

Continued next page

Crystal Creek Pizza Sauce:
1 (28-ounce) can diced tomatoes
2 roma tomatoes, chopped
1/2 cup chopped red onion
1/2 cup rehydrated chopped sun-dried
    tomatoes

4 cloves garlic, finely chopped
1 teaspoon olive oil
Salt and pepper to taste

Mix all ingredients together. Use a hand mixer or hand blender to smooth sauce, but don't puree.

Pizza Toppings: (Per pizza)
2/3 cup Crystal Creek Pizza sauce
8-10 thinly sliced pepperoni
4-6 thinly sliced red onion rings
1-2 sun-dried tomatoes, sliced thinly

1/2 orange or yellow bell pepper, cut into
    6-7 thin strips
1/4 cup grated Parmesan/Romano cheese
    blend
2/3 cup grated mozzarella cheese

Putting it together: Ladle sauce over pre-baked pizza crust. Evenly space pepperoni over sauce. Scatter onion rings, sun-dried tomato strips and bell pepper all over pepperoni. Cover with Parmesan/Romano and mozzarella cheeses. Bake at 425° until crust is golden brown and cheeses are bubbly.

Pollo e Pesce

The name of this dish is Italian for chicken with fish. It's a pounded chicken breast stuffed with spinach, proscuitto, wild mushrooms, sun-dried tomatoes and prawns. It's rolled, sautéed to a golden brown. To serve, you slice the chicken crosswise into 3 or 4 medallions and serve with a butter sauce. It's a beautiful and classic presentation, one I learned at culinary school. Chef Einck's version is just slightly different and incredibly elegant. My husband, who had never had it before, swears it's the best chicken dish he's ever eaten, which is a huge compliment, coming from an international business traveler who has eaten in some of the world's best restaurants.

Continued next page

Pollo e Pesce (continued)

1 SERVING

1 (8-ounce) boneless, skinless whole
    chicken breast
Flour for dusting
Salt and pepper
1-2 large leaves spinach
2 wild mushrooms, chopped (shiitake,
    oyster, cremini, etc.)
2 medium size shrimp, peeled, de-veined
    and butterflied
2 sun-dried tomatoes, cut into thin strips
2 thin slices of proscuitto
1 teaspoon olive oil

Sauce:
(Enough for 4 servings)
2 teaspoon olive oil
2 cloves garlic, finely chopped
1/2 cup white wine
2 sticks butter, cut into chunks
1 finely chopped seeded jalapeno
1/2 teaspoon chopped cilantro

Quickly blanch spinach by dropping into a pot of boiling water for 1 minute, then dropping into an ice bath for 1 minute to stop the cooking. Remove and lay on paper towels to dry. If using dried mushrooms and/or dried sun-dried tomatoes, rehydrate in boiled water for 20 minutes.

Pound chicken breast between 2 pieces of plastic wrap until an even 1/4" thickness. You will have a large, heart-shaped breast. Lightly dust with flour and clap the excess out with your hand. To construct, lay chicken flat with smooth side facing down and pointed end facing away from you. Season with salt and pepper. Cover breast with blanched spinach leaves. In the widest part towards you, lay the mushrooms, shrimp and sun-dried tomato. Cover the breast with proscuitto. Starting at the wide end, tightly roll the chicken toward the pointed end tucking in the sides as you go. Skewer securely. Preheat oven to 400°. Heat olive oil in a skillet over medium-high heat. Sauté chicken on all sides until golden brown, about 8 to 10 minutes. Finish cooking in a 400° oven for 10 to 15 minutes, or until chicken reaches 170°. Remove from oven and let rest 3 or 4 minutes before slicing into 3 or 4 medallions.

While chicken is finishing in oven, make the sauce. Heat 2 teaspoons olive oil in a sauté pan over medium-high heat. Add shallots then garlic. Cook until aroma is released, 1 minute or less. Add white wine and reduce to 1/4 cup. Remove from heat and whisk in a few chunks of butter, until melted, and then add a few more chunks. Repeat until all the butter is melted. You may have to put the pan back on low heat off and on during this process, but don't leave the pan on the heat for more than a minute at a time or the sauce will "break." Whisk in the jalapeno and cilantro and season with salt if necessary. Spoon warm sauce over chicken medallions and serve.

Classic Crème Brulee

Not all Crème Brulees are created equal. At least I've had a wide range of tastes and consistencies among the ones I've eaten over the years. I like a Crème Brulee that is light and creamy, almost like it's been whipped, but not quite that airy. That's how this recipe feels. The taste is rich with strong vanilla tones but not heavy like some I've tasted. Browning the sugar on top is a challenge with a conventional oven. It's much easier with a propane blowtorch. I think every cook should have a blowtorch in the kitchen. It's great for browning meringues as well as running off unwanted visitors. (Just kidding.)

6 SERVINGS

2 cups cream
1/2 vanilla bean, split
1/2 teaspoon vanilla extract

5 egg yolks
1/2 cup sugar
Sugar for dusting

Bring cream, vanilla bean and vanilla extract to a boil. Whip egg yolks and sugar until sugar is dissolved and mixture is thick and lemon colored. Pour a little hot cream into the eggs and stir, then add a little more and stir. Pour the tempered eggs back into the cream. Strain. Pour custard into 6 individual size ramekins. Set ramekins in a large roasting pan and pour in enough hot water to come half way up the ramekins. Bake at 300° for 30 to 40 minutes, or until custard is just set. Don't overbake. Remove from oven and roasting pan and cool until completely set and cold in the center. Sprinkle a thin layer of sugar over the top and place under a broiler, as close as you can get it, (or use a blow torch) until sugar browns, careful not to burn.

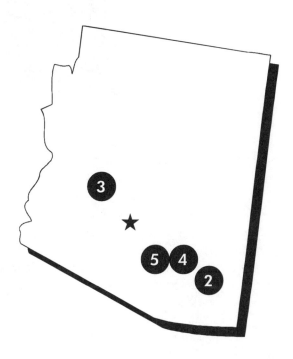

ARIZONA

Grapevine Canyon Ranch

P.O. Box 302
Pearce, AZ 85625
(520) 826-3185
Website: http://
    www.beadude.com

Season: Year-round

Capacity: 30

Accommodations: Casitas (suites with sitting area, bedroom and full bath); cabins (single room with bath); both have spacious decks and porches

Activities: Extensive horseback riding program; seasonal cattle work; sight-seeing of ghost towns/abandoned mining camps, historical fort and Indian burial grounds

Rates: $820-$945 per person, double occupancy per week; all meals and horseback riding included; other options available, call ranch for details, including daily and monthly rates

Grapevine Canyon Ranch, sequestered in the beautiful Dragoon Mountains in Southeastern Arizona, is a genuine working cattle ranch. Staff to guest ratio is 2 to 1, so count on plenty of attention and service. The guest program is almost entirely focused on riding. You can experience a wide range of riding programs at the ranch, including the exciting "adventure" rides, both on and off the ranch. We're talking serious remote backcountry rides, some lasting 5-7 hours at a time. The off-ranch excursions are awesome, including the famous Fort Bowie ride and the Chiricahua National Monument ride. Visiting these two sites alone is worth a trip to Arizona, but to do it via horseback, on little-traveled trails is quite another experience altogether.

Fort Bowie's history began in the 1860's when it was erected to protect the infamous Apache Pass. Most visitors have to walk a national forest service road 1½ miles to reach the fort. On horseback with a ranch guide, you see much more and truly get a feel for how our early pioneers traversed through this part of Apache country. Though this trip takes all day, all experience-level riders are welcome.

Chiricahua National Monument is perhaps the least known of all our national parks. In the 1920's, the park was established to protect the area for future generations. The scenery is breathtaking and this stunning canyon is known for its unusual and dramatic rock formations. Within the canyon itself you will experience several dramatically different landscapes, including soft, rolling meadows, steep mountains with tall pines and an almost lunar landscape at two of the more famous points of interest, Balancing Rock and Heart of Rocks. This ride is much more difficult than the Fort Bowie ride and may not be possible to make in the winter, as the elevation is 7,000 feet. (There is a small additional fee for the off-ranch horse trips.)

The ranch also offers what it calls "challenge" rides which are more difficult trail rides, and are as fast-paced as rider ability allows. Each rider is required to demonstrate his or her ability to the ranch wrangler in the arena before embarking on this Wild West jaunt. Of course, there are tamer trails for the novice rider and some excellent photographic rides. If you don't ride at all or want to take a day off, setting off for Old Mexico or the old west towns of Bisbee and Tombstone will make interesting day trips.

Regardless of the ride you take or the sightseeing you do, you'll want to come back to the ranch for a hearty Arizona-flavored meal. Every dish is made with fresh ingredients and prepared mostly from scratch. One thing for sure, there is always plenty of food and the kitchen staff is flattered when you go back for seconds. Because most of each day's activities involves hours of riding, most guests eat a hearty traditional style ranch breakfast. The breakfast menu consists of eggs, bacon, sausage, pancakes, omelets, biscuits and gravy, you name it. For a lighter appetite, the ranch offers a Grapevine specialty, Yogurt Parfait, which consists of layers of yogurt, fruit and granola in a pretty parfait glass. Dinners are chock full of protein and carbohydrates to replace the body's spent fuel and to recharge for the next day of saddle-hugging.

B.Hillis

Dinner Menu

GRAPEVINE BARBECUE RIBS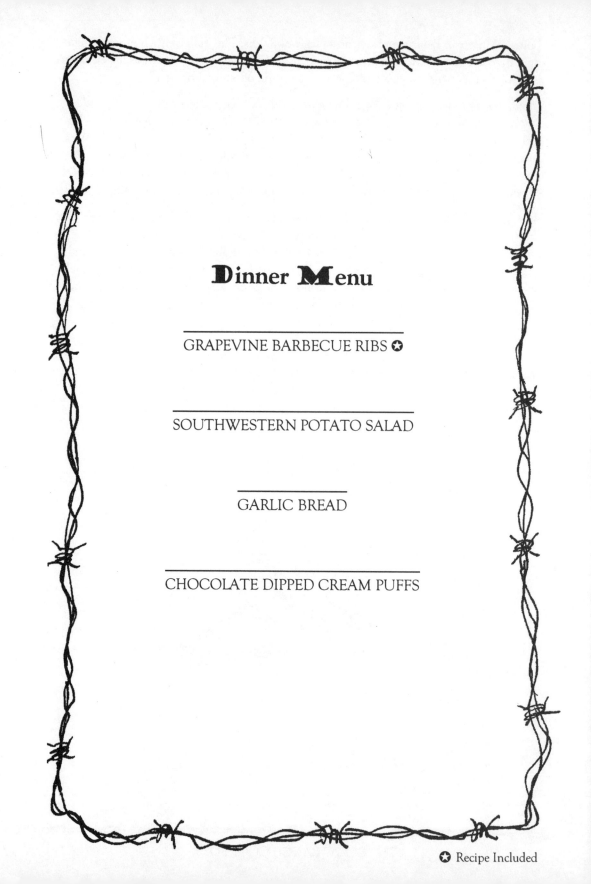

SOUTHWESTERN POTATO SALAD

GARLIC BREAD

CHOCOLATE DIPPED CREAM PUFFS

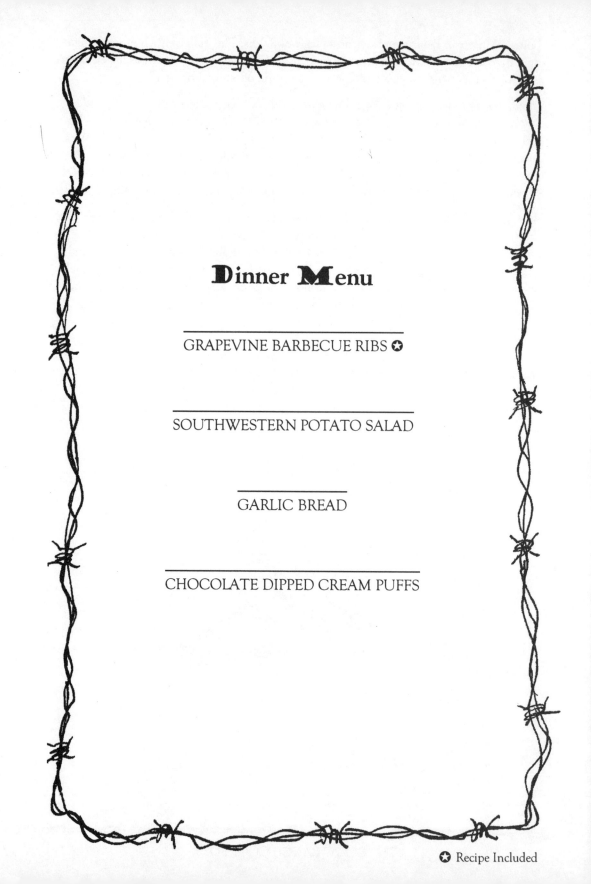 Recipe Included

Grapevine Barbecue Pork Ribs

You can't beat homemade barbecue sauces and this one's fantastic. It has a sweet smoky flavor and a great little kick of spice. It's an all-around good sauce to use on beef, pork or chicken. You'll find it tough to buy barbecue sauce at the grocery store after you've experienced this Arizona treat.

6 SERVINGS

6 to 8 pounds baby back pork ribs, trimmed
Garlic powder

Salt and pepper
3-1/2 cups Grapevine Barbecue Sauce, divided

Cut ribs into desired portion sizes (I like 4 to 5 ribs per section). Sprinkle both sides of ribs with garlic powder and salt and pepper. Place ribs in a large roasting pan and cover with foil. Bake at 275° for 2 to 3 hours, checking after 2 hours for tenderness. When desired tenderness is reached, remove ribs from oven and brush both sides with barbecue sauce (reserving some sauce to pass at service). Really slather the sauce on the ribs. Return to oven and cook another 1 to 1-1/2 hours uncovered. Heat the remaining sauce and pass with ribs.

Grapevine Barbecue Sauce:
1 (14-ounce) bottle ketchup
2 tablespoons tomato paste
3 ounces Worcestershire sauce
1/2 cup molasses
1-1/2 teaspoons liquid smoke (mesquite flavor preferably)
1/3 cup brown sugar

1/2 teaspoon dry mustard
1-1/2 teaspoons garlic powder
1/2 teaspoon cumin
1-1/2 teaspoons crushed red pepper
3/4 teaspoon cayenne
3/4 teaspoon chili powder
1/2 teaspoon ground coriander

Mix all ingredients together and heat until sugar dissolves.

Rancho de los Caballeros

1551 South Vulture Mine Road
Wickenburg, AZ 85390
(520) 684-5484

Season: October through May

Capacity: 175

Accommodations: Variety of casita style rooms and suites; including the charming original ranch rooms as well as newly constructed luxurious suites; all rooms appointed in beautiful southwestern décor with stunning desert views; Mobil Four Star property

Activities: horseback riding; top-rated 18-hole golf course; extensive children's program; cookouts; guest rodeos; swimming; tennis; nature hikes; trap and skeet shooting; conference facilities

Rates: Various options; golf or riding package (4 days/3 nights) $510 - $897 per person, includes all meals and lodging and greens fees or trail riding; non-package rates $178 - $499, includes all meals and lodging; no credit cards, please

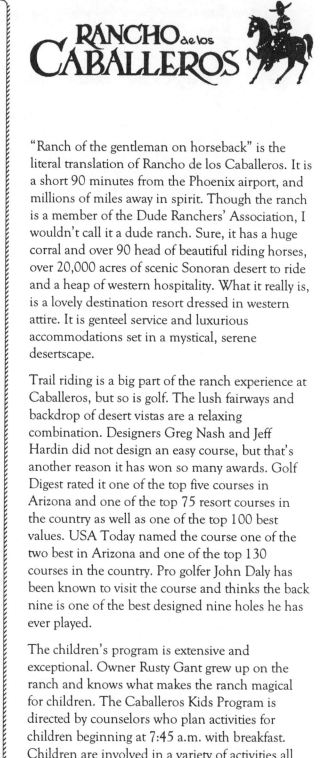

"Ranch of the gentleman on horseback" is the literal translation of Rancho de los Caballeros. It is a short 90 minutes from the Phoenix airport, and millions of miles away in spirit. Though the ranch is a member of the Dude Ranchers' Association, I wouldn't call it a dude ranch. Sure, it has a huge corral and over 90 head of beautiful riding horses, over 20,000 acres of scenic Sonoran desert to ride and a heap of western hospitality. What it really is, is a lovely destination resort dressed in western attire. It is genteel service and luxurious accommodations set in a mystical, serene desertscape.

Trail riding is a big part of the ranch experience at Caballeros, but so is golf. The lush fairways and backdrop of desert vistas are a relaxing combination. Designers Greg Nash and Jeff Hardin did not design an easy course, but that's another reason it has won so many awards. Golf Digest rated it one of the top five courses in Arizona and one of the top 75 resort courses in the country as well as one of the top 100 best values. USA Today named the course one of the two best in Arizona and one of the top 130 courses in the country. Pro golfer John Daly has been known to visit the course and thinks the back nine is one of the best designed nine holes he has ever played.

The children's program is extensive and exceptional. Owner Rusty Gant grew up on the ranch and knows what makes the ranch magical for children. The Caballeros Kids Program is directed by counselors who plan activities for children beginning at 7:45 a.m. with breakfast. Children are involved in a variety of activities all morning and rejoin their parents after lunch. From 6:00 p.m. to 9:00 p.m., children are welcome to join ranch counselors for an evening of fun and active events such as scavenger hunts, marshmallow roasts and talent shows. The children's program is steeped in tradition, as is the entire ranch program.

The ranch opened its doors 50 years ago and is still owned and operated by the same family. Dallas "Rusty" Gant's father started the ranch with two other partners in 1947. The first guests arrived in December 1948, and have been streaming in ever since.

Rusty and his staff, led by Cathy Weiss Pereira, maintain the old charm of the original ranch while inconspicuously upgrading the service and amenities. Cathy's background in luxury hotel management serves the Caballeros guests well. Every evening, Cathy makes a point of visiting with each and every guest to ascertain his or her satisfaction. She does it with a sincere heart and a genuine love for the guests and the ranch she serves.

The ranch was and still is a marvelous piece of architecture, including its famous hand-dug, pearl-shaped swimming pool, quite a daring undertaking for its day. The living room is a comfortable and cozy gathering place, dominated by a copper-hooded fireplace stamped with the Caballeros brand, and the historic Santa Fe furniture designed by the late artisan Bruce Cooper. There is a certain nostalgic charm to the main lodge and the guest quarters. If you turned around and saw Cary Grant leaning on the bar, you wouldn't be surprised. (He and other Hollywood stars have graced the doors of the ranch over the years.) You feel like you've stepped back in time the minute you arrive at the ranch.

The ranch is only 2 miles from the small, historic town of Wickenburg, once a booming mine town. The town is home to the famous Desert Caballeros Western Museum, housing an impressive western art collection featuring Remington, Russell and Moran. Just a few minutes from the ranch is the abandoned site of the fabled Vulture Mine, which produced over $200 million in gold during its heyday. The mine shafts, buildings and living quarters look like the miners just walked away. As wind whistles through the saguaros, one gets an eerie feeling walking through the property. A lonely doll lies lifeless on a child's bed, a mallet sits idle on a work table, and dishes wait patiently on the kitchen counter. It makes you wonder why those early dwellers left the camp so strangely intact.

Return to the ranch after a day of sightseeing, golfing or horseback riding and you will have worked up an appetite worthy of hearty ranch cuisine. Chef Dan Martin has been overseeing the culinary operation at the ranch for 20 years. His signature style blends the flavors of the southwest with old-fashioned comfort food and exquisite presentation. Five entrees are featured nightly. You might see slow-roasted rack of pork or braised lamb with wild mushrooms baked in a puff pastry. Apple Almond Chicken is a guest favorite as are the southwestern signature dishes. And just when you thought your evening couldn't end more perfectly, you're tempted with an array of fabulous desserts from Mary Jane Almand, who has been delighting guests for 22 years with her delectable pies, pastries, breads and desserts. A true story, recently a guest staying in Wickenburg was so distraught over that particular establishment's dessert, Mary Jane was called upon to provide one of her famous pies to save the day.

Breakfast is a feast to behold and served in the sunny enclosed patio room beside the pearl pool. Six or seven fresh fruit choices, along with an assortment of breads and pastries and a cold cereal/granola/yogurt bar are tempting until you see the waffles or pancakes or biscuits and gravy. My personal favorite is a made-to-order omelet prepared by Chef Martin himself, with a cornucopia of toppings from which to choose. The service is impeccable at all meal periods with polite servers quietly in and out at your table, anticipating your every need.

Chef Martin and Mary Jane have shared some of their favorite recipes with us and all of them embody the spirit of Rancho de los Caballeros. I hope you enjoy them as much as I have.

THE GREAT RANCH COOKBOOK

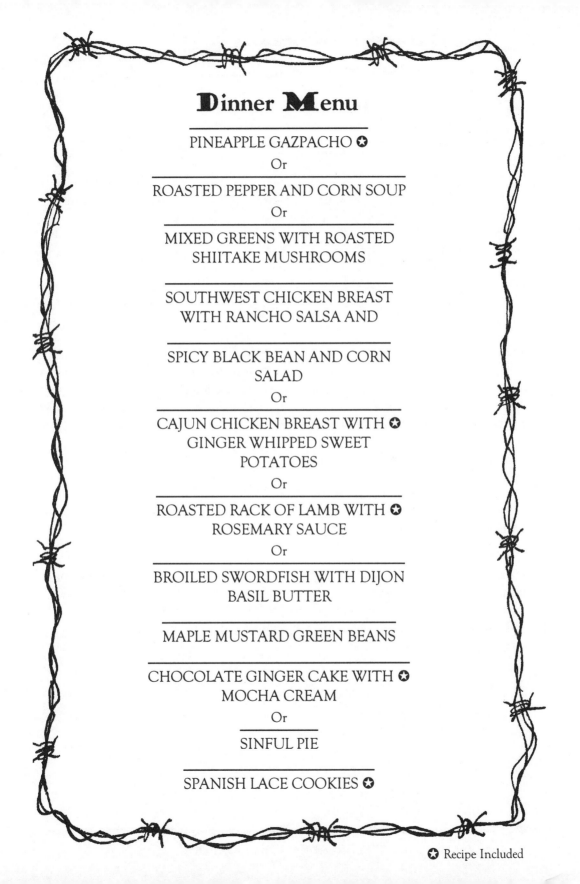

Dinner Menu

PINEAPPLE GAZPACHO ✪

Or

ROASTED PEPPER AND CORN SOUP

Or

MIXED GREENS WITH ROASTED
SHIITAKE MUSHROOMS

SOUTHWEST CHICKEN BREAST
WITH RANCHO SALSA AND

SPICY BLACK BEAN AND CORN
SALAD

Or

CAJUN CHICKEN BREAST WITH ✪
GINGER WHIPPED SWEET
POTATOES

Or

ROASTED RACK OF LAMB WITH ✪
ROSEMARY SAUCE

Or

BROILED SWORDFISH WITH DIJON
BASIL BUTTER

MAPLE MUSTARD GREEN BEANS

CHOCOLATE GINGER CAKE WITH ✪
MOCHA CREAM

Or

SINFUL PIE

SPANISH LACE COOKIES ✪

✪ Recipe Included

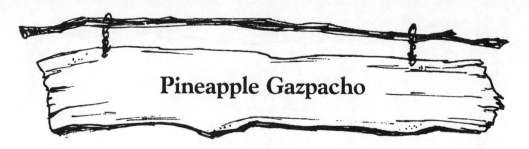

Pineapple Gazpacho

The answer is yes. This does taste as interesting as it sounds. What a unique idea, too. Great for those parched, sun-drenched southwestern summer days, this soup is delightfully refreshing. Just so you don't forget that it is a southwestern dish, there is a nice little heat kick in the aftertaste. It's a lovely icy-pale yellow color with splashes of red and green peppers. Make the soup the day before you plan to serve it to give the flavors time to harmonize.

4 SERVINGS

1/2 cup chopped yellow bell peppers
1/4 cup pineapple juice
1 (1-1/2 pound) fresh pineapple, chopped
1/4 cup chopped red onion
1/4 cup peeled, seeded and chopped
    cucumber
2 teaspoons rice wine vinegar
1/4 teaspoon salt
1/2 teaspoon hot red pepper sauce
1/8 teaspoon white pepper
1 tablespoon brown sugar.

Garnish:
1/4 cup finely chopped red bell pepper
1/4 cup finely chopped green bell pepper
1/4 cup finely seeded, chopped cucumber
1 tablespoon chopped cilantro

Place first 9 ingredients (yellow bell pepper through white pepper) in a food processor or blender and puree. Pour into a mixing bowl and add brown sugar and stir. Refrigerate (can be prepared 1 day in advance). To serve, ladle 6 ounces into a chilled soup cup and sprinkle with red and green bell peppers, cucumber and cilantro.

Cajun Chicken Breast

Now this is spicy hot! I've provided two blackening seasoning recipes (1 hot and 1 less hot) but you could use a store-bought Cajun seasoning mix. The only thing I don't like about the store brands is the enormous amount of salt. Chef Martin pairs this with the Ginger Whipped Sweet Potatoes which almost cool and certainly sooth the spiciness of the chicken.

4 SERVINGS

4 boneless, skinless chicken breasts
1/2 cup Cajun seasoning mix (recipe follows)
2 tablespoons clarified butter* or vegetable oil

Preheat oven to 350°. Heat a cast iron skillet over medium heat. While skillet is heating, coat one or both sides of the chicken with the seasoning mix. Add clarified butter or oil to the skillet and cook the chicken breast until crust forms on both sides, about 3 to 5 minutes per side. Finish in a 350° oven, 5 to 10 minutes or until done.

Very Spicy Blackening Seasoning:
2 tablespoons paprika
1 tablespoon chili powder
1 tablespoon garlic powder
1 tablespoon ground Mexican oregano
1 tablespoon onion powder
1-1/2 teaspoons cayenne
1-1/2 teaspoons white pepper
1-1/2 teaspoons black pepper
1-1/2 teaspoons salt

Mix all ingredients and store in an airtight container.

Spicy Blackening Seasoning:
3 tablespoons paprika
1 tablespoon chili powder
1 tablespoon garlic powder
1 tablespoon ground Mexican oregano
1 tablespoon onion powder
1 teaspoon cayenne
1 teaspoon white pepper
1 teaspoon black pepper
1 teaspoon salt

Mix all ingredients and store in an airtight container.

To clarify butter, see page 12.

Ginger Whipped Sweet Potatoes

Ginger and sweet potatoes really go together well. The whipped consistency
is lovely and I can't think of a better dish to pair with the spicy Cajun
Chicken Breast. I like these just the way they are, but my tester Kathy thought
a tablespoon of brown sugar enhanced the ginger flavor.

4 SERVINGS

3 sweet potatoes, peeled and cut into
    chunks
2 tablespoons fresh grated ginger

1/4 cup butter (1/2 stick), cut into chunks
Salt and white pepper to taste

Boil sweet potatoes until very tender (about 20 to 25 minutes) and drain. With an
electric mixer, mix the potatoes, ginger and butter until light, smooth and fluffy. Add
salt and white pepper to taste.

Spanish Lace Cookies

These cookies have made Chef Mary Jane famous at the ranch. They are beautiful
and look like lace because they are filled with holes. These cookies are delicate
but crunchy and taste buttery. Try a teaspoon of cinnamon for a variation.

YIELD = 48 (3") COOKIES

1/2 cup light corn syrup
1/2 cup butter (1 stick)
2/3 cup brown sugar

1 cup flour
1 cup chopped pecans

Preheat oven to 350°. In a medium-size saucepan, bring the corn syrup, butter and
brown sugar to a boil, stirring occasionally. Add the flour and pecans and stir until well
mixed. Return to just a boil (1 to 2 minutes). Remove from heat and set aside to cool for
about 10 minutes. Make teaspoon-size balls and place on a cookie sheet lined with
parchment paper. Bake 8 to 10 minutes or until a rich golden brown color. Don't let it
get too dark. Remove from oven and cool.

Roasted Rack of Lamb
with Rosemary Sauce

Authorities on a good lamb dish, Art and Bonnie Cikens tasted these chops side by side with the chops from Elk Creek. We had a split decision. Art preferred this rich, brown rosemary sauce while Bonnie preferred the Elk Creek fruited sauce. Both are delicious. I'd have to say that this rosemary dish is more traditional. The sauce is smooth, and the lemony-pine flavor of the rosemary really comes through.

4 SERVINGS

4 lamb racks (4 ribs each), trimmed and cleaned

3 cloves garlic, finely chopped
Salt and pepper to taste

Preheat the oven to 350°. Rub the racks with garlic and season with salt and pepper. Heat a large oven-proof skillet over medium high heat. Sear racks on all sides in hot skillet. Place skillet in oven and finish cooking lamb. It will take about 15 to 20 minutes to reach 130°, medium-rare. Remove from oven when desired temperature is reached and let rest, covered loosely with foil for 5 to 7 minutes before carving.

Rosemary Sauce

YIELD = 1-1/2 CUPS

2 cups beef or lamb stock
1 tablespoon butter
1/4 cup flour

3 tablespoons fresh rosemary leaves
3 cloves garlic, finely chopped
Salt and pepper to taste

In a large saucepan, heat stock. In a small saucepan, heat butter and add flour, stirring to make a roux. Cook the roux for 5 to 10 minutes, or until it smells nutty and looks golden. Add roux to heated stock. Add rosemary and garlic and lower heat. Cook for 10 to 15 minutes. Strain and season with salt and pepper. Thin with more heated stock if necessary.

Note: We also made this sauce with the Demi-Glace Gold® from More Than Gourmet. We used 2 (1.5 ounce) tins, reconstituted them according to the directions and eliminated the roux from this recipe. To order Demi-Glace Gold®, see More Than Gourmet in the "Sources" section on page 236.

Chocolate Ginger Cake
with Mocha Cream

Mmmm — a moist, chocolate gingerbread cake. The coffee-flavored frosting
is light and airy and complements the cake nicely. It's not an overly sweet
cake, even with the frosting. It doesn't look like it's loaded with a lot of fat,
either. Dare I say it's low-fat and spoil your fun?

9 SERVINGS

1-1/4 cups brown sugar
2/3 cup buttermilk
1/2 cup applesauce
1/4 cup vegetable oil
2 eggs
1 teaspoon vanilla
1 cup flour

2/3 cup cocoa powder
1 teaspoon baking soda
1/2 teaspoon baking powder
1/2 teaspoon salt
1 tablespoon ground cinnamon
1 tablespoon fresh grated ginger

Preheat oven to 350°. In a large bowl, mix the brown sugar, buttermilk, applesauce, oil,
eggs and vanilla. In another bowl, mix the flour, cocoa powder, baking soda, baking
powder, salt and cinnamon. Sift the dry mixture into the liquid mixture and whip until
smooth and add ginger and mix well. Pour batter into a greased 8" X 8" baking dish and
bake for 30 to 45 minutes, or until a toothpick inserted in the center comes out clean.
Remove from oven and cool in pan 10 minutes. Remove from pan and cool on wire
rack. When cool, frost with Mocha Cream (recipe follows).

Mocha Cream

1-1/2 cups heavy cream
2 tablespoons powdered sugar

1 tablespoon instant coffee granules
3 tablespoons cocoa powder

Whip heavy cream on medium speed for 1 or 2 minutes. Add powdered sugar, coffee
and cocoa and whip until stiff peaks form, about 2 minutes more. Spread evenly over
cooled chocolate ginger cake. Dust with a little cocoa powder.

Tanque Verde Ranch

14301 E. Speedway
Tucson, AZ 85748
(520) 296-6275
E-mail: dude@tvgr.com

Season: Year-round

Capacity: up to 200

Accommodations: Spacious patio casitas with mountain views and southwestern decor; Mobil Four Star Ranch

Activities: Arizona's largest riding stable; mountain biking; hiking programs; nature programs and museum; evening programs; children's programs; 5 tennis courts, indoor/outdoor pools, saunas; whirlpool; exercise room; corporate/business retreat

Rates: winter $290-$380, summer $200-$265 (per night/double occupancy)

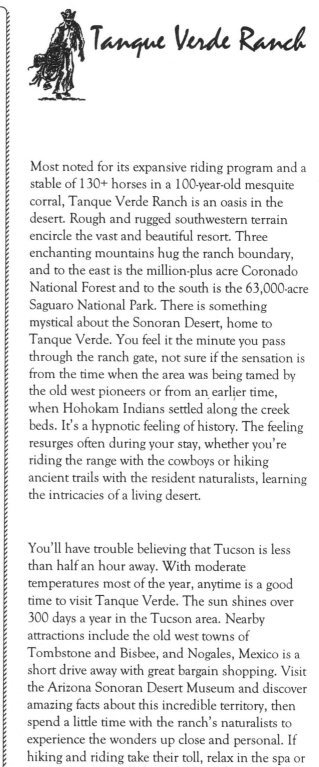

Most noted for its expansive riding program and a stable of 130+ horses in a 100-year-old mesquite corral, Tanque Verde Ranch is an oasis in the desert. Rough and rugged southwestern terrain encircle the vast and beautiful resort. Three enchanting mountains hug the ranch boundary, and to the east is the million-plus acre Coronado National Forest and to the south is the 63,000-acre Saguaro National Park. There is something mystical about the Sonoran Desert, home to Tanque Verde. You feel it the minute you pass through the ranch gate, not sure if the sensation is from the time when the area was being tamed by the old west pioneers or from an earlier time, when Hohokam Indians settled along the creek beds. It's a hypnotic feeling of history. The feeling resurges often during your stay, whether you're riding the range with the cowboys or hiking ancient trails with the resident naturalists, learning the intricacies of a living desert.

You'll have trouble believing that Tucson is less than half an hour away. With moderate temperatures most of the year, anytime is a good time to visit Tanque Verde. The sun shines over 300 days a year in the Tucson area. Nearby attractions include the old west towns of Tombstone and Bisbee, and Nogales, Mexico is a short drive away with great bargain shopping. Visit the Arizona Sonoran Desert Museum and discover amazing facts about this incredible territory, then spend a little time with the ranch's naturalists to experience the wonders up close and personal. If hiking and riding take their toll, relax in the spa or

take a refreshing dip in one of 3 sparkling pools. Don't miss the sweeping sunsets from the comfort of your patio or the porch of the main lodge. It's likely you've not seen such a display of brilliant oranges, fiery reds and hot yellows, or cool blues, delicate pinks and soft purples. The dramatic show is brief, but don't despair if you miss it. It plays again tomorrow, and the next day and the next.

Tanque Verde Ranch is also a working cattle ranch, with more than 340 head of cattle. Guests are welcome to participate in the 3 annual cattle round-ups. The ranch is also noted for its extensive children's program and, of course, for its eloquent cuisine. It's not only the guests who want to stay forever at the ranch. The staff does, too. Executive Chef Mark Shelton has been at Tanque Verde Ranch since 1980! Most of his staff have been at the ranch in excess of 5 years, giving guests the consistency in excellence they deserve.

The menu changes daily and guests have been known to write home about the lunch dessert table — 16 feet of homemade cakes, pastries and pies. In addition to continental and American West cuisine worthy of a Mobil Four Star property, guests also enjoy outdoor barbecues with a roaming country and western singer, mountain breakfast rides and all-day pack trips with savory surprises from Chef Shelton's kitchen. Back at the ranch, the guests settle into the beautiful southwestern dining room, featuring 3 fireplaces, a 16-foot saguaro-ribbed ceiling and thick, massive walls. Artfully decorated with original western art and authentic Navajo weavings, the alluring ambiance is outdone only by the feast of flavors created by Chef Shelton and his skilled staff.

Chef Shelton has shared a lovely breakfast menu and two outstanding dinner combinations, showing the range of talent and breadth of flavors that would fulfill any serious gourmand's fantasy.

B.Hillis

Breakfast Menu

FRESH FRUIT CUP ✪
WITH MESQUITE HONEY

ORANGE CREAM CHEESE ✪
STUFFED FRENCH TOAST WITH
BLUEBERRY SYRUP

THICK SLICED
MESQUITE-SMOKED BACON

HONEY-GLAZED BANANA MUFFINS

Dinner Menu

REFRIED BEAN SOUP

SESAME ENCRUSTED PORK TENDERLOIN ✪
WITH MANGO BERRY SALSA

Or

SOUTHWESTERN VEAL CHOP WITH ✪
TOMATO GREEN CHILE SAUCE

CHIPOLTE GOUDA SMASHED POTATOES ✪

BLACKBERRY PIE ✪

✪ Recipe Included

Orange Cream Cheese Stuffed French Toast

I do love the blueberry syrup with this toast, but it masks the orange flavor somewhat. If you want a more pronounced orange flavor, serve this with pure maple syrup. Try it with the blueberry syrup first to see how delicious it is.

4 SERVINGS

2 (8-ounce) packages cream cheese, softened
1/2 cup orange marmalade
4 eggs, beaten
1 tablespoon sugar

1/4 teaspoon vanilla
Pinch nutmeg
8 (1-1/2") slices day old French bread
1 tablespoon butter
2 cups Blueberry Syrup (recipe follows)

Whip cream cheese with marmalade until smooth and creamy. With a sharp knife, cut a slit (2-1/2"-3") lengthwise through the bottom crust of the bread to create a pocket for stuffing. Run the knife back and forth to loose the inside as close to the edges you can get without cutting through the bread. Load a 12" or large pastry bag fitted with a flat tip with the cream cheese mixture. Squeeze about 2 or 3 tablespoons of filling in each slice, making sure the filling is distributed evenly within the bread slice. In a large bowl, beat eggs, sugar, vanilla and nutmeg until well-mixed. Dip each side of the stuffed bread into the batter and set on a sheet pan. Melt butter in a skillet or griddle over medium heat. When hot, add bread and cook until golden brown on both sides, about 4-5 minutes each side. Serve warm with blueberry syrup (recipe follows).

Blueberry Syrup

A more appropriate title might be blueberry sauce. This is fairly thick to be called a syrup. It reminds me more of a thick fruit glaze. Using arrowroot instead of cornstarch will provide for a less starchy-tasting syrup.

YIELD = 2 CUPS

2 cups frozen blueberries, divided
1/2 cup sugar
1/2 cup light corn syrup

1/2 cup water
2 tablespoons arrowroot or cornstarch
2 tablespoons water

Mix 1 cup of blueberries, sugar, corn syrup and 1/2 cup of water in a saucepan and bring to a boil. In a small dish, mix arrowroot with 2 tablespoons of water until smooth. Pour into boiling blueberries and return to a boil. Stir in remaining cup of blueberries and cook 2 minutes. Thin with water if too thick. Can be served hot or cold.

Fresh Fruit Cup with Mesquite Honey

People say that mesquite honey tastes different from regular clover honey. I guess all honey tastes a little different from the next one. If you can't find mesquite honey, use your favorite and this dish will still taste good. Not only is it pretty, it tastes light and luscious. Use only fresh berries, so if you can't locate one or two of them, just make the dish with what you can find.

6 SERVINGS

1/2 cantaloupe	1/2 cup raspberries
1/2 honeydew	1/2 cup blueberries
1/2 pineapple	1/2 cup blackberries
1-1/2 cups grapes	1/2 cup mesquite honey
1/2 cup strawberries	2 tablespoons brandy (preferably Calvados)

Cut cantaloupe, honeydew and pineapple into bite-size pieces. In a large bowl, mix cut fruit with grapes. In a small bowl mix honey and brandy. Pour half of honey mixture over cut fruit and stir. Scoop individual portions of glazed fruit into small bowls. Cut strawberries in half. Garnish fruit bowls with strawberries, raspberries, blueberries, blackberries and a sprig of mint. Drizzle with remaining honey mixture.

B. Hillis

Sesame Pork Tenderloin

Fruit salsas are awesome when they accompany a warm meat or fish. This berry salsa really bursts with flavor and makes the pork oh so much better than by itself. The sesame seeds add a nice nutty taste so you get multiple flavor explosions all at the same time.

4 SERVINGS

2 pork tenderloins (1 pound each)
Salt and pepper to taste

1/2 cup sesame seeds
1 tablespoon sesame oil

Trim silverskin from pork if necessary. Season with salt and pepper. Roll pork in sesame seeds. Heat a large sauté pan on medium-high heat. Add sesame oil and sear pork on all sides until browned and semi-crisp. Finish in a 375° oven until temperature reaches 150°. Let rest 5 minutes, covered loosely with foil. Slice on the bias and serve with fruit salsa.

Mango Berry Salsa

1 large mango, peeled, seeded and finely chopped
1/4 cup raspberries
1/4 cup blackberries
1/4 cup blueberries
2 tablespoons finely chopped mint

Mix all ingredients and let stand a room temperature until ready to serve.

Southwestern Veal Chops
with Tomato Green Chili Sauce

This is a very innovative southwestern dish. The sauce is superb, smoky and chile-flavored. My tester Daniel made this dish and my husband was the official taster. I've never seen two men dive into food the way they attacked this veal. Nothing was left on the plate but cleaned bones. It's one of those dishes that makes you smile because you know your guests really loved it. I think that's a big part of the reason we cook.

4 SERVINGS

4 (10 to 12 ounce) cleaned, trimmed veal chops
2 cloves garlic, finely chopped
1/2 teaspoon charbroil or mesquite seasoning
1/2 teaspoon paprika
1/4 teaspoon salt
1/4 teaspoon pepper

Sauce:
1 poblano chile, roasted*
2 cups tomato sauce
2 cups finely chopped fresh tomatoes
1/2 teaspoon ground cumin
1/2 teaspoon ground coriander
1 teaspoon chili powder
Salt and pepper to taste

Preheat oven to 400°. In a small bowl, mix the garlic and seasonings and rub on both sides of the veal chops. Heat a skillet over medium-high heat. When hot, sear chops until brown on both sides. Finish in the oven until chops reach 140°-145°. Remove and let rest, loosely covered with foil for 5 to 7 minutes. Serve with Tomato Green Chili Sauce.

To make sauce, simmer tomato sauce over medium heat for 10 minutes. Add tomatoes and poblano chile. Simmer 15 minutes or until sauce is slightly thickened. Season with cumin, coriander, chili powder and salt and pepper.

*To roast the poblano pepper, see page 12.

Chipolte Gouda Smashed Potatoes

Just when you thought there were no more ways
to serve mashed potatoes. Voila!

Give me a spoon and a pan of these potatoes and call me happy.
The chipolte provides a friendly punch while the
Gouda gives a smooth nutty flavor.

4-6 SERVINGS

5 baking potatoes, peeled and quartered
1/2 cup sour cream
1 cup milk

1 cup shredded Gouda
1 chipolte pepper, finely chopped*
Salt and pepper

Boil potatoes until very soft, about 25 to 30 minutes. Drain liquid and add sour cream, milk, cheese and chipolte. Mash until almost smooth but still some lumps. Season with salt and pepper.

*Chipolte pepper is a smoked jalapeno. It is available in a dried form from Pecos Valley Spice Company, or in a can with Adobo Sauce from A.J.'s Fine Foods. If you cannot find it at your store, check out one of these sources. See the "Sources" section on page 236. If you use the dried form, reconstitute the chile in boiling water with a teaspoon of vinegar first.

Blackberry Pie

This is a custard-based fruit pie. I thought that was kind of unusual but it works really well. The berries are tart by nature and the egg custard tames the edgy berries with a smooth silky flavor. The dough recipe is enough for 2 double-crusted pies. Use one set now and freeze the other.

YIELD - A 9" PIE

Pie Dough:
4-1/2 cups cake flour
1-1/3 cups bread flour
1 tablespoon salt
2-1/3 cups shortening
3/4 to 1 cup milk

Filling:
2 eggs
1 cup sugar
1 tablespoon butter, melted
1/4 teaspoon nutmeg
1/4 cup all-purpose flour
4 cups fresh blackberries

To make crusts, mix cake and bread flours with salt. Cut shortening into flour mixture. Add just enough milk to make dough come together. Turn out onto a lightly floured surface. Knead 2 or 3 times, just enough to blend dough. Divide into 4 equal balls. Wrap 2 for the freezer and chill the other two, wrapped, in the refrigerator for 30 minutes before rolling. When thoroughly chilled, roll 1 crust about 1/4" larger than the pie pan. Center in pie pan and press dough onto bottom and sides. Refrigerate while you make filling.

To make filling, beat eggs and sugar until smooth. Add nutmeg, butter and flour. Mix well. Place fruit evenly over bottom pie shell. Pour custard over fruit. Roll top crust 1" larger than pie pan. Center over top of pie. Tuck top crust edge under bottom crust and crimp with a fork or flute with your fingers to seal crust. Cut 4 slits in the center of the pie for venting. Bake at 400° for 15 minutes, then reduce heat to 350° and bake another 40 to 50 minutes. You may need to cover piecrust edges with foil if it's getting too brown. Remove from oven and cool at least 20 minutes before serving.

White Stallion Ranch

9251 W. Twin Peaks Road
Tucson, AZ 85743
(520) 297-0252

Season: September through May

Capacity: 50-75

Accommodations: Spanish-style bungalows with white adobe exteriors; single rooms, suites and deluxe suites; private baths, air-conditioning; most have private patios

Activities: Horseback riding; team penning; hayrides; cookouts; nature walks; astronomy program; hiking; swimming; tennis

Rates: $230-$350 per couple per night

Year after year, third and fourth generations of guests continue to trek back to the True family's White Stallion Ranch in the southwestern desert. Bringing the children and the children's children seems as natural as putting your boots on, one foot at a time. The hospitality at White Stallion is genuine, and the Trues know what it means to have family gathered 'round. The third generation of Trues is growing up on the ranch today.

Just outside of Tucson, the White Stallion sprawls over 3,000 acres of unspoiled desert landscape. When Cynthia and Allen True bought the ranch in 1965, it was only 160 acres. At the time, there were 30 guest ranches within an hour of Tucson. Now there are 3. Urban growth was killing the ranches by developing the open land. In an effort to preserve the ranch life for their own family as well as others, the Trues began purchasing land around the ranch soon after they arrived. Rest assured that the tranquil, traditional way of ranch life will continue at White Stallion for generations to come.

The pinnacle of White Stallion is, as the name suggests, horsemanship. Not that you'll find a white horse (they don't do well in the Arizona sun), but you will find a full stable of gentle, trained riding and roping horses. There are breakfast rides, mountain rides, all-day rides and sunset rides. Gather down at the corral and participate in a guest/ranchhand rodeo, including the team penning. With the help of the ranch's experienced cutting horses, guests and ranch employees team up to separate a young calf from

the herd and wrangle it into a pen. It's a popular western rodeo event and is much more challenging than it might sound. Those little dogies have a staunch instinct and powerful determination to get back to the herd. It's great fun and most guests ask to do it again and again during their stay.

A stay at White Stallion is good for the soul, and wonderful for bringing families closer together. The ambiance is leisurely, the staff is as friendly and warm as family, the scenery is spiritually soothing and the food is comforting and delicious. Most of the meals are served family style and the recipes have been in the family for generations. You are likely to see plenty of grilled steaks and chicken and all the traditional fixings like baked and fried potatoes, corn-on-the-cob and cowboy beans. Lunches are tasty casseroles like sour cream enchiladas, Reuben casserole and carrot quiche. Breakfasts are hearty and include traditional bacon and eggs, pancakes and waffles as well as some southwestern specialty dishes like breakfast burritos and huevos rancheros. My all-time favorite has to be the dessert packed in a lunch box for an all-day ride. These peanut butter bars can be dangerous, though, because you will want to eat them all in one sitting.

Peanut Butter Bars

It's amazing what these little bars can do. My neighbor, Kim Boerner, was adamant about not tasting any of the "treats" I was conjuring up during this cookbook testing. She showed unbelievable willpower — until I swirled a plate of these under her nose. She caved in. Rosalie, who tasted many of the recipes for this book, asked if I could put a heart beside this recipe so you would know how wonderful they are. She's always thinking of others. The bars are chewy and the topping is silky. The creamy peanut butter flavor comes through first, then you get the chocolate flavor. In a word — yummy!

YIELD = 24 BARS

1/2 cup shortening
1/4 cup butter (1/2 stick)
3/4 cup peanut butter
1 cup brown sugar
1 egg
1 teaspoon vanilla
1-1/4 cup all-purpose flour
1-1/4 cup oatmeal
3/4 teaspoon baking soda
1/4 teaspoon salt
1 cup semi-sweet morsels

Topping:
1 cup powdered sugar
1/2 cup peanut butter
1 teaspoon vanilla
2 to 5 tablespoons milk

Preheat oven to 350°. Beat the shortening, butter, peanut butter, brown sugar, egg and vanilla in a large bowl until smooth. In a separate bowl, mix the flour, oatmeal, baking soda and salt. Stir the flour mixture into the peanut butter mixture until well incorporated. Spread the dough by hand in a 9" X 13" lightly greased baking dish. Bake 15 to 17 minutes or until the edges just barely start to turn brown. The middle will look like it is not done, but you want to slightly under-cook these so that they will be chewy. Turn the oven off. Immediately sprinkle the semi-sweet morsels evenly over the dish. Return the pan to the oven for 1 or 2 minutes. Remove from the oven and spread the melted morsels evenly over the top.

To make the topping, beat all the ingredients together until smooth, starting with 2 tablespoons of milk and adding more to reach a creamy consistency that is easy to spread. Swirl the topping over the melted chocolate. You'll get an almost marble-like look from the chocolate and topping blending together. Let cool then cut into 24 bars.

CALIFORNIA

The Alisal Guest Ranch and Resort

1054 Alisal Road
Solvang, CA 93463
(805) 688-6411
(800) 4-ALISAL
Fax (805) 688-2510
Website: http://www.alisal.com
E-mail: info@alisal.com

Season: Year-round

Capacity: 150-200

Accommodations: 73 guest cottages, configured in either comfortable studios or spacious two-room suites. Lodging includes a full breakfast and lavish dinner.

Activities: Horseback riding; 2 golf courses; 7 tennis courts; fishing (instruction for additional fee); hiking; bicycling; boating; nature walks; a host of children's activities; wine country tours; group and corporate special activities.

Rates: 2-night minimum stay; $335-$415 per night, double occupancy, depending upon room; includes breakfast and dinner; ask about the Round-Up vacation package.

THE ALISAL

You would never, in a million years, imagine that the Alisal Guest Ranch is a working cattle ranch. It's too luxurious to conjure up the image of dusty trails and sedulous cattle branding. The history of the ranch is fascinating, as it was one of four original land grants on the West Coast given to conquistador Raimundo Carrillo in the late 1700's for his service to the newly formed Mexican government. A world-class resort, the 10,000-acre Alisal is among the West's most sought-after destinations not only for individuals but also for group meetings and corporate retreats. With a repeat rate of 80%, it's no wonder. Secluded deep in the Santa Ynez Valley, between Santa Barbara and Santa Maria, the Alisal offers about every amenity and activity you could possibly want. Its "Round-Up" vacation package offers golf, tennis, horseback riding and fishing. If those activities aren't enough, there are also bicycling, bird watching, hiking and local art galleries, museums and shopping.

One of the most enticing draws to the Alisal is its location amid the world-renowned wine country, Santa Barbara County. There are more than 30 wineries within driving distance of the ranch and most are open for tours and tasting. The wineries often host winemaker dinners at the ranch.

Dining at the Alisal is a treat all of its own, with incredible breakfast buffets as well as a la carte selections that would enlighten even the finickiest breakfast eater. Tables brimming with fresh fruit, homemade baked breads and pastries and esculent entrees beckon you away from your comfortable

garden-encased cottage. Daily breakfast features Alisal's famous buttermilk pancakes, Huevos Rancheros and the Vegetarian Breakfast Burrito, among other things. You could stay at the Alisal for weeks and never get bored with the selections.

Dinners are elegant and bountiful. It's a time for dressing up and enjoying a leisurely evening, reminiscent of European-paced dinners. The Alisal features a rotating menu, with several delectable options each night. Fortunately for you, and the Alisal, is the presence of Executive Chef Pascal Gode (boy, I do love his accent!) who trained in his native France starting at the tender age of 13. His career has been studded with star restaurants (including his own) and resorts such as the Adolphus Hotel, the Four Seasons Hotel and Resort and others. The food at the Alisal is as good as it gets. Start with either gourmet soup or a fresh pasta creation or a choice of a specialty salad or mixed greens salad. There are 8 different entrees from which to choose, which might feature Black Angus beef, free range poultry, fresh fish from the Pacific and a "traditional" ranch selection that is anything but "conventional." The sauces alone tempt the soul of any worthy gourmand, e.g., Cabernet Sauvignon-Maui Onion Butter, Saffron Beurre Blanc, Gingered Plum sauce, and a sweet, smoky Mole Poblano, to name a few. Finish the exquisite meal with one of the ranch's homemade desserts such as Apple Cobbler with Cinnamon Ice Cream, Peanut Butter Pie, or the much requested Ranch Oatmeal Pie. If you have room, a double espresso or a Cappuccino crowns the evening and sends you off to another blissful night, dreaming about tomorrow's adventure.

John Martino, Alisal's Food and Beverage Director, has shared a typical menu with us, and included the ranch's most requested recipes:

Alisal's Most Requested Recipes

BUTTERMILK PANCAKES

TORTILLA SOUP

BRAISED LAMB SHANKS WITH A
RED WINE SAUCE

PEANUT BUTTER PIE

Monday's at Alisal's

Choice of one
Soup of the day

Linguine Pasta with Toasted Pumpkin Seeds and Ancho Chili Sauce

Greek Salad with Feta Cheese and Kalamata Olives
Lemon Herb Dressing

Mixed Greens Salad – Chopped Scallions, Parmesan Cheese,
Croutons and Choice of Dressing

Entrees
Grilled Angus "Cowboy Cut" Rib-Eye Steak
With Yucca Fries and Garlic Herb Butter

Curry-Rubbed Turkey Paillard
With Homemade Mango Chutney

Chicken and Shrimp Fricassee "Louisiana Style"
Served with Pan-Seared Corncakes

Fresh Fish of the Day

Oven-baked Potato with Summer Vegetable Medley

Traditional Ranch Entrée
Rock Sugar Cured Pork Loin
With Caramelized Onion and Apple Stuffing

Alisal BBQ Special
Braised Smokey Lamb Shoulder Chops
With Stewed Tomatoes and Buttered Egg Noodles

Alisal Buttermilk Pancakes

These are the lightest, fluffiest pancakes I've ever tasted. The buttermilk adds a nice tang, too. They don't even need butter and are perfect with just a splash of pure maple syrup or whipped cream and strawberries.

2-4 SERVINGS

1 egg
1-1/4 cups buttermilk
2 tablespoons oil
1 teaspoon vanilla extract
1 cup all-purpose flour
1 teaspoon sugar
2 teaspoons baking powder
1/2 teaspoon baking soda
1/2 teaspoon salt

In a small mixing bowl, beat the egg with a fork. Beat in buttermilk, oil and vanilla. Set aside. In a large mixing bowl, mix the flour, sugar, baking powder, soda and salt. Add egg mixture to dry ingredients. Stir the mixture until blended but slightly lumpy. Lightly grease a griddle or heavy skillet and heat to medium low or medium. Pour 1/4 cup of batter for each pancake. Cook 2 or 3 minutes, until bubbles form and pop on surface and edges start to dry. Flip pancake over and cook another 2 to 3 minutes. Serve immediately. Makes 8 or 9 (4") pancakes.

Tortilla Soup

I know why this is one of the Alisal's most requested recipes. The broth is so flavorful it's like biting into a fresh homegrown tomato. My tasters practically licked the bowl clean. If you like a little more "heat" don't remove the seeds from the jalapeno.

6 SERVINGS

2 tablespoons peanut oil
2 each, corn tortillas cut into 1" squares
2 tablespoons chopped garlic
1/2 onion, chopped
1 jalapeno pepper, seeds and veins
    removed
2 quarts chicken stock*
1/8 cup tomato paste
3 teaspoons ground cumin
2 cups chopped tomatoes

Garnish:
2 cups cooked and cubed chicken breast,
    about 4 breasts**
2 cups diced avocado
1 cup grated cheddar cheese
1/2 cup chopped cilantro
4 corn tortillas, cut into thin strips and
    fried crisp in 1/4 cup oil

Prepare all the garnishes. Cook corn tortilla 1" squares in oil until crisp. In a food processor, mix the garlic, onions, jalapeno and tomatoes by pulsing 10 times or continuously for 10 seconds. It will look a little like mush. Add this mixture to stockpot with tortilla squares. Add tomato paste and cumin. Sauté 4 to 5 minutes. Add chicken stock and simmer until liquid is reduced by 1/3, about 20 to 30 minutes. Strain and adjust seasonings with salt, black pepper and cumin. To serve, ladle 6 to 8 ounces of soup into bowl and top with chicken, avocado, cheese, cilantro and fried tortilla strips.

Concentrated chicken stock is available through More Than Gourmet. See "Sources" section on page 236 for details. One (1.0-ounce) tin of Fond de Poulet Gold® makes 5 cups of chicken stock. You will need 8 cups for this recipe.

**My good friend Chef Donna Bachman taught me the perfect way to cook chicken breasts without drying them out. Place the breasts on a baking sheet with at least 1/2" sides, pour half & half over them and bake at 350° for 30 minutes. You won't believe how moist the chicken is.*

Braised Lamb Shanks

The lamb meat is so tender after simmering in a lush red wine sauce it falls off the bone. If you like mint jelly with leg of lamb, you'll want to have some for this dish, too. The sauce by itself is enough to make this dish. It's so flavorful and intense. You will swear there is cream in it but there isn't.

6 SERVINGS

6 lamb shanks
1/2 cup flour for dredging
1/2 cup olive oil, divided
1 medium onion, finely chopped
1 bay leaf
2 small carrots, finely chopped
1 stalk celery, finely chopped

3 cups dry red wine, about
    1 (750 ml) bottle
2-1/2 cups chopped roma tomatoes
1 teaspoon tomato paste
1-1/2 tablespoons chopped parsley
2 cloves garlic, finely chopped
1 tablespoon finely chopped orange peel
Salt and pepper to taste

Dredge the lamb shanks in flour. Heat a sauté pan over medium high heat. Brown in a pan with all but 2 tablespoons of olive oil. Remove the shanks and keep warm.

Add the remaining 2 tablespoons of olive oil to the pan and sauté the onion, carrot, celery and bay leaf. Cook 5 minutes and add the wine. Simmer until most of the wine has evaporated from the pan, about 15 minutes.

Add the lamb shanks to the pan with the tomatoes and tomato paste. Cover and simmer on low heat 1-1/2 hours or until tender.

Remove the shanks from the pan and strain the sauce. Add the garlic, parsley and orange zest into the sauce. To serve, spoon sauce over shanks and sprinkle with a little parsley.

Alisal's Peanut Butter Pie

Peanut butter and chocolate were made for each other. This pie is as terrific a marriage between the two as you'll find. The filling is so creamy and the chocolate topping is just the right amount to balance the flavors. This is sure to become a family favorite.

YIELD = 1 PIE

Crust:
1-1/3 cups graham cracker crumbs
1/3 cup sugar
8 tablespoons butter, melted

Filling:
12 ounces cream cheese at room
    temperature
1-1/2 cups peanut butter
1-1/2 cups sugar
1 cup heavy cream

Topping:
1/2 cup sugar
1/2 cup heavy cream
2 ounces bittersweet chocolate, chopped
1/2 teaspoon butter
1/2 teaspoon vanilla extract

Preheat oven to 350°. Prepare crust by combining graham cracker crumbs, sugar and melted butter in a mixing bowl. Press the mixture in the bottom and sides of a 10" pie pan. The crust will shrink as you bake it, so push it up as high as you can on the sides. Bake the crust for 8 to 10 minutes. Set aside to cool.

Prepare the filling by mixing the cream cheese, peanut butter and sugar until well-blended. Whip the heavy cream until stiff. Fold it into the cream cheese mixture. Spoon the filling into the cooled crust.

Prepare the topping by combining the sugar and the cream in a saucepan. Bring to a boil then reduce heat and simmer without stirring for 6 minutes. Remove from heat and add the chocolate and butter and stir until melted. Stir in vanilla. Carefully pour the topping over the pie, and refrigerate uncovered for at least 4 hours.

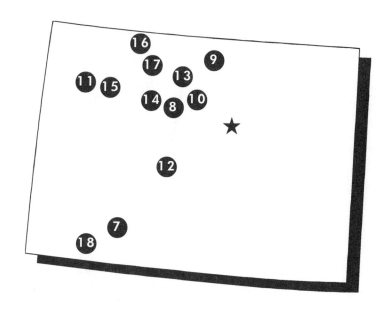

COLORADO

4 UR Guest Ranch

P.O. Box 340
Creede, CO 81130
(719) 658-2202

Season: June through September

Capacity: 50

Accommodations: 3 separate lodges with 5 to 8 rooms each, 1 guest cottage for 4-6 people

Activities: Fly-fishing, hot springs; horseback riding; hiking; jeep tours; trap shooting; tennis; swimming; picnics and cookouts; children's program

Rates: $175-$195 per person per day ($150 per child); July and August 1 week minimum stay; special rates for conferences, seminars and special group meetings

Southwestern Colorado is home to the history-rich 4 UR Ranch, nestled in the Goose Creek Valley of the San Juan Mountains. The hot springs on the property attracted early-day entrepreneurs and curiosity-seekers, and by the late 1800's a hotel was built touting the medicinal and restorative powers of the 30 bubbling pools. At the turn of the century, the man who founded the towns of Durango, Alamosa and Colorado Springs bought the ranch and opened his own guest ranch for friends and business associates. The 4UR Ranch has been opening its valley and its doors to guests for more than 100 years. The ranch is an authentic old west destination with all the modern conveniences for a relaxing and comforting vacation.

The fishing is superb largely due to an aggressive water management program employed by the ranch to protect the pristine waters and fish population, and the environmentally sensitive guests the ranch attracts. Excellent fishing awaits you on Goose Creek and the Rio Grande. For a special fly-fishing adventure, embark on a guided trip to the Lost Lakes, two crystal clear alpine lakes with incredible fishing and breathtaking views.

Take a break from the fishing and saddle up for some terrific trails accompanied by an experienced wrangler. The new Junior Wrangler program is designed to teach children all about horses and riding, leaving them with an appreciation for the gentle creatures and their importance in ranch life. Even if you've never ridden a horse, the patient wranglers (and horses) will have you riding on the open range before the end of your stay.

The hot springs, while purported to have medicinal properties, will at least relax and refresh your body and soul. Other activities are available to fill your days such as swimming, tennis, rafting and hiking. Listen for the dinner bell and head to the main lodge for some fantastic western cuisine. Superb food prepared by a seasoned Chef and kitchen staff awaits your arrival in the sun porch dining room. Gorgeous views of the surrounding mountains and glimpses of steam rising off the hot sulfur pool remind you that you're far away from the hustle and bustle of city life. There is an early continental breakfast for the fishers followed by a full course breakfast headlining special dishes daily. Don't miss the weekly breakfast ride featuring biscuits and gravy, scrambled eggs and ham, baked apples and cowboy coffee. Every Friday brings a high-noon fish fry along Goose Creek. Or try a backcountry gourmet picnic. Dinners are elegant and creatively presented. Choose between a couple of different entrees nightly.

B.Hillis

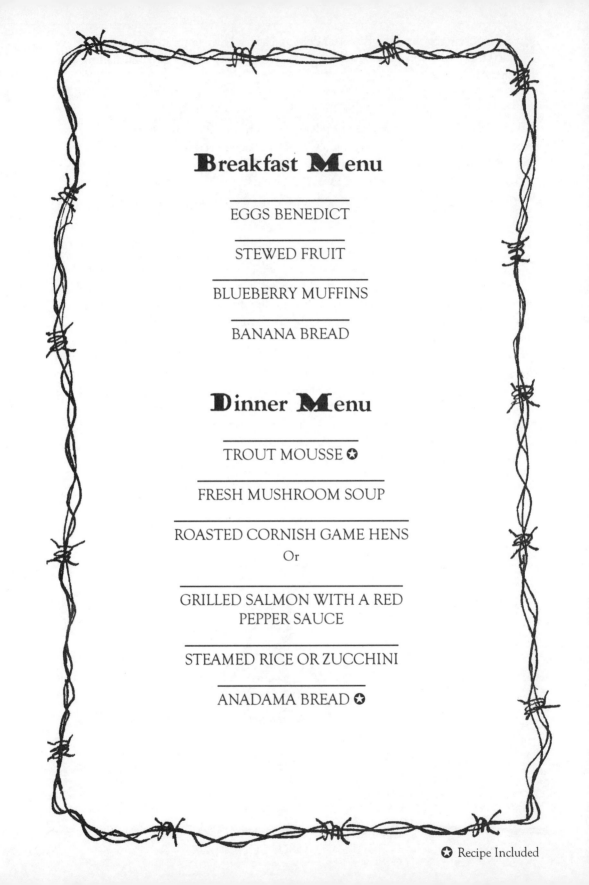

Breakfast Menu

EGGS BENEDICT

STEWED FRUIT

BLUEBERRY MUFFINS

BANANA BREAD

Dinner Menu

TROUT MOUSSE ✪

FRESH MUSHROOM SOUP

ROASTED CORNISH GAME HENS
Or

GRILLED SALMON WITH A RED
PEPPER SAUCE

STEAMED RICE OR ZUCCHINI

ANADAMA BREAD ✪

✪ Recipe Included

Trout Mousse

When I first saw the recipe name, I thought, "that's kind of odd." Now I think, "that's pretty cool." It's different and tasty and easy to make. My husband, who loves trout any which way (especially catching them, and then letting them go) thought this mousse was delicious. It's a keeper.

6 SERVINGS

3 (10") fresh trout, heads and tails removed
2 quarts water
1 rib celery, finely chopped
1/2 bunch green onions, finely chopped
1/2 tablespoon whole black peppercorns
1/4 cup lemon juice

1 pound cream cheese, softened
2 tablespoons lemon juice
1 teaspoon liquid smoke
1/4 teaspoon salt
1/4 teaspoon pepper
1 tablespoon fresh chopped dill
1 tablespoon fresh chopped chives
2 cloves garlic, finely chopped
1 tablespoon finely chopped red onion

Bring water, celery, green onions, peppercorns and 1/4 cup lemon juice to a simmer. Add fish and simmer for 15 minutes. Remove fish and cool for 30 minutes. Meanwhile, with a mixer, whip cream cheese until light and fluffy, about 5 minutes. Add the rest of the ingredients (except the fish) and whip for another 2 minutes. Scrape down sides of bowl and mix 1 more minute. Make sure there are no lumps. If there are lumps, mix again.

Remove the fish meat from the bones and add meat to cream cheese mixture. Mix on medium speed for 1 minute. Taste and adjust seasonings if necessary with salt and pepper. Refrigerate, covered, until needed. Keeps in the refrigerator for up to 1 week. Serve with crackers or toasted baguette slices.

Anadama Bread

Legend has it that a farmer invented this colonial bread. Apparently, his wife left him with nothing but a bit of molasses, cornmeal and flour. If that is true, I'm impressed that he knew what to do with them. Sometimes I just don't give men enough credit. I had never tasted this bread before, so I can't compare it to other Anadama recipes. All I can say is that it's pretty darn delicious and makes your kitchen smell good while it's baking. If you've never had it either, expect a dense, sweet and dark amber-colored loaf. It's really good served warm with butter or honey butter.

YIELD = 2 LOAVES

2 cups milk
1/2 cup yellow cornmeal
2 teaspoons salt
1/2 cup molasses
3 tablespoons oil

1/3 cup warm water
4-1/2 teaspoons yeast (2 (.25-ounce) packages)
5 cups all-purpose flour, more or less

Combine milk, cornmeal and salt in a medium saucepan. Heat to boiling, stirring constantly. Reduce heat to low and cook 5 minutes, stirring occasionally. (The mixture will be really thick, like mush.) Add molasses and oil and stir. Remove from heat and cool to lukewarm.

In a large bowl, add warm (110°) water and sprinkle with yeast. Let sit for 2 minutes. Add cornmeal mixture to yeast and stir well. Add 2 cups of flour and mix until all flour is incorporated. Add enough remaining flour to make a stiff dough. Let the dough rest for 10 minutes. Mix in mixer with a dough hook on medium-slow speed for 5-7 minutes or until dough is smooth and elastic. (You can knead by hand for 10 minutes if you prefer.) Place dough in a greased bowl, turning once to bring up greased side. Cover with plastic wrap then a kitchen towel and set in a warm place to rise until doubled in size, about 40-60 minutes.

Without punching down, turn dough out on lightly floured surface. Divide in half and shape into 2 loaves. Place in greased 9" X 5" pans. Cover and let rise again until doubled, about 40 minutes. (Dough will be about 1" above top of pan.) Bake in a preheated oven at 350° for 40 to 50 minutes. Turn out of pans and cool on rack.

Aspen Canyon Ranch

13206 County Road #3
Parshall, CO 80468
(970) 725-3600
Website: http://
    www.pageplus.com/export/
    DUDE/aspen_canyon.html
E-mail: tiger@colorado.net

Season: Summer June through
October; winter snowmobiling
tours mid-December through
April; fall hunting October
through November

Capacity: 40

Accommodations: Three four-
plex log cabins, nestled along the
Williams Fork River. All quarters
have natural gas fireplaces,
carpeting, porches; the Cliff
House, a 3 bedroom/3 bath
home available for large parties/
families

Activities: fishing; hiking; riding;
roping; pack trips; white-water
rafting; rodeos, children's rodeo

Rates: $1,195 adults, $650
children, per person weekly

ASPEN CANYON RANCH

Tucked serenely away in the Williams Fork Valley
is the quaint, beautiful Aspen Canyon Ranch.
Only 90 miles west of Denver in reality, it feels
like a million miles in perception. "A river runs
through it" is an accurate description, as the
Williams Fork River meanders through the
property as does Lost Creek, both brimming with
native trout. The guest cabins are all built as close
to the river as possible, so upon rising in the
morning, you can see and hear the gurgling water
swirling past your cabin and think about the
brook, brown or rainbow trout waiting to catch
your fly. The ranch covers nearly 3,000 acres and
has a full cattle operation. You can participate in
the ranch work if you like, and depending upon
the time of year you visit, you might be mending
fences, rounding up cattle or harvesting the hay.

Aspen Canyon is a family ranch, geared toward
making sure everyone has a good time, from the
wee ones to mom and dad. The extensive
children's program allows parents to do their own
exploring without worrying about the kids. You
can be sure they are having the time of their life. It
will be difficult to decide which activities to choose
to fill your days because there are so many. Fishing
both on the ranch and off is some of the best
Colorado has to offer. A rafting trip on the Upper
Colorado River is a refreshing break from
"cowboying." Miles of hiking trails await your
footsteps and of course horseback riding with your
personal wrangler is a must. You can learn how to
rope if you'd like, or even race barrels.

One thing not to miss if you do ride is the all-day ride to Williams Peak, overlooking the Gore Mountain range. At 11,000 feet, the view is incredible. You wonder how anything could be so gorgeous. You tell your mind to etch the view permanently so that weeks from now, when you're back up to your eyeballs in stress you can close your eyes, and take a deep breath, and transport yourself back to that peak. That special moment, standing on Williams Peak, soaking up the view, is so rare that once you experience it, you feel blessed. Few places on earth can compare to the majestic beauty of Williams Peak.

If you want to explore nearby attractions, Breckenridge, Vail, Winter Park and the Rocky Mountain National Park are all within a short drive. The ranch can also, with advanced notice, schedule other activities for you, like mountain biking, backpacking or even hot air ballooning. Great service, delicious western cuisine and tons of activities await you in this quiet little valley surrounded by majestic mountains.

Speaking of outstanding food, the culinary experience at Aspen Canyon is reason enough to book a trip. Dine in a relaxed family atmosphere where everything you taste is homemade, including breads, desserts, soups and even the chocolate chip cookies found in every guestroom. Julie Hoffman, who has been cooking at the ranch for over 10 years, heads the kitchen staff. She has perfected the art of western cuisine with flair and style and keeps her staff focused on the mission of serving quality fare that keeps the guests coming back for more. The week is speckled with traditional cookouts and there is a fabulous early morning breakfast ride and a dinner ride you won't want to miss. Julie sets out on the trail ahead of you and begins preparing a spread of food that you'll likely smell before you can see it. The ranch chose a trout dinner to share with us because it's not the kind of meal we would eat on a regular basis. The breakfast is a favorite among returning guests and more than likely, it will be one of your favorites, too.

Breakfast Menu

YOGURT FRUIT CUP ✪

BAKED FRENCH TOAST WITH ✪
RASPBERRY SYRUP

LINK SAUSAGE

GRANOLA MUFFINS

Dinner Menu

MARINATED ASPARAGUS SALAD

ASPEN CANYON RANCH BREAD ✪
WITH HONEY BUTTER

BAKED TROUT WITH ✪
PEPPERS AND ONIONS

STEWED APPLES

WILD RICE

CHOCOLATE NUT TORTE

✪ Recipe Included

Yogurt Fruit Cup

My neighbors Jan and Pat Johnston helped me taste this lovely concoction. I think they were relieved I didn't show up with another plate of muffins or bags of granola. Jan thought it was a great dish because the dressing wasn't too sweet and it didn't overpower the fruit flavors. The combination of fruit is excellent, too. You could use whatever fruit you like, but we think the ranch picked out a very appropriate combination that provides ample contrast in both flavor and texture. It's even kind of healthy, too, if you're into that sort of thing.

6-8 SERVINGS

1 cup sliced peaches, unpeeled
1 cup grapes, sliced in half (any kind of seedless grape)
1 cup chopped cantaloupe
1 cup fresh blueberries

1 cup plain yogurt
1-1/2 + teaspoons lemon juice
1/2 + teaspoon poppy seeds
1 tablespoon honey

Combine all the fruit in a large bowl. Mix the yogurt, lemon juice, poppy seeds and honey together and gently fold into the fruit. Serve immediately.

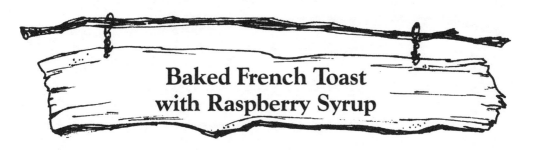

Baked French Toast with Raspberry Syrup

I really like the fact that you prepare this dish the night before, so you just pop it in the oven in the morning and — voila! — instant breakfast. I like the taste even better than the ease of preparation. It's crunchy on the outside (the kind of crunch you can't get with pan-fried French toast) and soft on the inside. I could eat this toast all by itself (it's that good), but then I'd miss the luscious Raspberry Syrup and how it enhances the warm toast. The recipe calls for white bread. I used the Aspen Canyon Ranch Bread I had made the day before. That might be the reason this dish was so very good.

4 SERVINGS

8 slices firm white bread
2 tablespoons sugar
1/4 teaspoon baking powder
3 eggs
1-1/2 teaspoon vanilla
1-1/2 teaspoon fresh grated orange peel

1/2 teaspoon cinnamon
3/4 cup milk
1/4 cup orange juice
1/4 cup powdered sugar
Fresh raspberries for garnish (optional)

Lightly spray a 9" X 13" baking dish with non-stick cooking spray. Cut bread slices in half diagonally, and arrange them in the baking dish in overlapping rows. Mix together the sugar and baking powder. Whisk in the eggs, vanilla, orange peel and cinnamon. Gradually whisk in the milk and orange juice. Pour egg mixture evenly over bread. Cover dish with plastic wrap. Place a second baking dish on top to weigh down the toast and refrigerate overnight. In the morning, uncover French toast and bake at 375° (starting with a cold oven, so you don't crack your dish) for 25 to 35 minutes, or until it is puffed and well-browned on top. Dust with powdered sugar and serve immediately with warm Raspberry Syrup (recipe follows) and garnish with fresh raspberries if in season.

Raspberry Syrup:
3/4 cup maple syrup

1/2 teaspoon cinnamon
1-1/4 cups raspberries, divided

In a small saucepan, heat half of the raspberries with the maple syrup and cinnamon. Cook over medium heat until syrup is almost boiling, about 5 minutes. Remove from heat and stir in the remaining half of raspberries. Serve warm. You can store this syrup in the refrigerator for 2 days.

Aspen Canyon Ranch Bread

This is one of those breads you could sit down and eat the whole loaf right out of the oven. It's yeasty, soft, and tender. The honey butter is a nice touch, too. There are 3 risings to this bread, which is why it's so soft and fluffy.

YIELD = 1 LOAF

1 tablespoon yeast
1 cup warm water (110°)
1 teaspoon sugar
3 cups all-purpose flour

2 tablespoons oil
1 teaspoon salt
4-1/2 teaspoons sugar

In a large mixer with a dough hook, dissolve yeast in warm water, then add 1 teaspoon sugar and stir. Set aside for 3 to 5 minutes, or until mixture becomes bubbly. In a separate bowl, mix flour, salt and sugar. Add flour mixture and oil to yeast mixture. Mix on low speed until dough forms a ball. Turn mixer to medium speed and mix for 4 minutes, or until dough is smooth and elastic. Place dough in a large greased bowl, turning once to coat both sides with oil. Cover with plastic wrap, then a towel and place in a warm place to rise.

When dough has doubled in size (about 45 to 60 minutes), turn out on a lightly flour-dusted surface and punch down. Put back in the bowl and allow it to rise a second time, covered, for about 35 to 45 minutes. Punch down again and form into a rectangle loaf to fit a greased 9" X 5" loaf pan. Cover pan with plastic wrap and allow bread to rise a third and final time until it rises about 1"-2" above top of pan. Preheat oven to 375° and bake for 45 to 55 minutes, or until top is golden brown and loaf sounds hollow when thumped.

Honey Butter:
1/4 pound butter (1 stick) softened   1/4 cup honey

Mix butter and honey in a mixer until smooth. Pack into a decorative bowl or ramekin and chill. Let sit at room temperature 30 minutes prior to serving.

Rainbow Trout
with Peppers and Onions

I normally don't like to look my food in the eye when I'm eating it. It didn't bother me to devour this moist, flaky fish and it didn't bother our golden retriever Georgia either as she "fished" the skeleton out of the trash when I wasn't looking. Bad dog! I can't really blame her; this is one tasty trout.

5 SERVINGS

5 fresh whole Rainbow trout (cleaned with head and tail intact)
2 tablespoons olive oil
Salt and pepper

1 red bell pepper, cut into strips
1 green bell pepper, cut into strips
1/2 cup sliced onion (half the onion lengthwise first, then cut into strips)
1 tablespoon olive oil

Preheat oven to 350°. Lay trout on a lightly greased or parchment-lined baking sheet. Brush trout inside and out with 2 tablespoons olive oil. Sprinkle salt and pepper on the inside of the trout. Sauté red and green peppers and onions in 1 tablespoon of olive oil until tender-crisp, about 2 or 3 minutes. Place the sautéed vegetables inside the trout, dividing the mixture evenly. Bake for 15 to 20 minutes or until flesh is flaky. Serve immediately.

Aspen Lodge
at Estes Park

6120 Highway 7
Longs Peak Route
Estes Park, CO 80517
(970) 586-8133
Website: http://
www.aspenlodge.com

Season: Year-round

Capacity: 150

Accommodations: 36 rooms in the
main lodge; 24 multi-room cabins
with porches and mountain views

Activities: Summer — horseback
riding; hiking; fishing; swimming;
tennis; sports center; golf nearby;
square dancing; live
entertainment; Winter —
snowmobiling; cross-country
skiing; ice skating; snowshoeing;
sleigh rides; horseback riding

Rates: Summer - 3, 4 or 7 night
packages (7 nights $880 per adult,
all meals and lodging); winter -
$79-$89 per night for 2 people,
breakfast included

Recently named "Best Dude Ranch" by the Official
Hotel Guide Travel Agent Subscribers, Aspen Lodge
is located in north central Colorado, in the midst of
beautiful Rocky Mountain territory. Only seven miles
from downtown Estes Park, the lodge is far enough
away to feel like you're isolated from the bustling pace
of city life, but close enough for a quick trip to town
for incomparable shopping at over 300 stores and
galleries. The lodge, Colorado's largest log building,
is nestled within 80 acres of open meadows at a
staggering 9,000-ft. elevation. Aspen Lodge also leases
a 3,000 acre ranch that borders the Rocky Mountain
National Park and Roosevelt National Forest, giving
you more than 33,000 acres of riding and hiking
trails. It would be tough to hit them all in just one
trip.

Open all year, the lodge has activities for everyone,
especially the children. The lodge has developed a
children's program that combines fun with education
and exploration. Children will learn about early
Indian life, pioneer life-styles and the nature and
environment in northern Colorado. For adults, in
addition to riding and hiking, you can try mountain
biking and fly-fishing in the lodge's own lake or in
nearby waters. Winter brings a whole array of
outdoor activities, including snowshoeing, skiing or
snowmobiling on those same beautiful summer trails.
Though wildlife viewing is abundant all year, it's
especially exciting to see elk and deer in the pristine
winter wonderland.

You'll also be pleased with the cuisine, ranging from
traditional western to gourmet continental. I like the
lodge's use of wild game in a tame setting. The recipe
Aspen Lodge shares with us is a great example of how
it blends local game with classic techniques to create a
uniquely Colorado cuisine.

Breakfast Menu

FRUIT AND YOGURT

THE ASPEN LODGE SKILLET

MULTI-GRAIN PANCAKES

APPLE CINNAMON TURNOVERS
WITH VANILLA GLAZE

Dinner Menu

GRILLED SHRIMP BROCHETTES

COLORADO ELK MAISON ✪

GARLIC HERBED MASHED
POTATOES

SAUTÉED VEGETABLES WITH
ANCHO CHILE-ARTICHOKE RELISH

TIRAMISU

✪ Recipe Included

Colorado Elk Maison

Wow! I had never had elk before, so I wasn't sure what to expect. The meat is darker than beef, and leaner. The taste is stronger than beef but not as strong as venison, and it tastes best at medium rare. Even if you don't like elk or you can't find it, try this delectable sauce on beef tenderloin or even pork tenderloin. It has a deep earthy flavor and coats your mouth with a lovely Madeira aftertaste. I was able to locate elk tenderloin through the help of the incredibly resourceful butcher for A.J.'s Fine Foods, Bob Stachel. His number is in the "Sources" section on page 236. Tell him I sent you.

6 SERVINGS

2-1/2 pounds elk tenderloin
Salt and pepper
Sauce:
1 teaspoon finely chopped shallots
4 tablespoons vegetable oil

1 large portabella mushroom cap, sliced
1/2 cup Madeira wine
1-1/2 cups heavy cream
2 tablespoons chopped parsley
Salt and pepper

To prepare the sauce, heat oil in a 2-quart saucepan over medium-high heat. Sauté shallots for 1 or 2 minutes. Pull pan away from flame and add Madeira (careful, the wine is highly flammable). Cook until the wine is reduced by half then add mushrooms and heavy cream. Bring to a boil and reduce heat to a steady simmer. Cook until sauce thickens and mushrooms are tender, about 15 to 20 minutes. The sauce should coat the back of a spoon without dripping. Add salt and pepper to taste.

Slice elk into 12 equal size medallions. Heat grill to medium high heat (375°-400°). Grill medallions to desired doneness, about 2 or 3 minutes per side for medium rare. Arrange 2 medallions on a warm plate. Spoon mushroom sauce on lower 1/3 of medallions and on plate in front of medallions. Sprinkle with parsley.

C Lazy U Ranch

P.O. Box 379
Granby, CO 80446
(970) 887-3344
E-mail: clazyu1946@aol.com

Season: June through September,
mid-December through March

Capacity: 110

Accommodations: 41 individual
units, all with private bath and
75% have fireplaces; no phones
or TV; all units meet the AAA
Five Diamond and Mobil Five
Star standards, but vary in size
and décor

Activities: Summer — horseback
riding; fly-fishing; hiking; tennis;
swimming; western dancing;
children's program; cookouts;
wagon rides; skeet and trap
shooting;
Fall — adult only format;
horsemanship clinics; horseback
riding; fly-fishing; hiking; skeet
and trap shooting; Winter —
cross-country, downhill and
Telemark skiing; sledding; winter
indoor and outdoor horseback
riding; ice skating; children's
program; holiday festivities

Rates: Average weekly per person
(call ranch for children's rates)
Summer and fall- $1,700
Holiday - $1,525
Winter - $120-$205 per night, 2
night minimum; no credit cards

C Lazy U is the ONLY guest ranch in the U.S. to
ever receive the AAA's Five Diamond award and
the Mobil Five-Star resort award. Each year the
ranch continues to maintain the high standards
required by these two prestigious travel
organizations and capture these awards, both of
which are based on a total package of service,
amenities and overall atmosphere.

Vacationing at C Lazy U seems like visiting
another world, one filled with happy
accommodating staff, tons of fun activities,
beautiful scenery and unbelievable food. C Lazy U
is snuggled in the Willow Creek Valley,
overlooking the Continental Divide, about 2 hours
northwest of Denver. The ranch sits on 2,000
acres with access to 3,000 more. Your most
difficult decision will be when to go. The summers
are awesome but the winters are incredible. Toss a
coin, or better yet, experience both seasons.

The ranch is owned and operated by the Murray
family, a close-knit group actively involved in
helping guests have the vacation of their dreams.
While the summertime is a great time to visit and
the riding program and other activities are
outstanding, the fall and winter programs
contribute to C Lazy U's uniqueness. In the fall,
the ranch hosts horsemanship and cow-working
clinics. The clinics are designed to improve your
skills in managing and riding horses and give you
a ranch vacation at the same time. You can bring
your own horse, or borrow a ranch horse, but
having your own horse, whether you bring it or
not, is usually required. You learn to work with a
horse's spirit, instead of against it and how to earn
the horse's trust and build confidence between you
and the horse. Graduates of the horsemanship
program can be considered for enrollment in the
cow-working clinic, which takes horsemanship to a

whole new level. Both are phenomenal programs and best of all, woven into an overall fantastic ranch vacation. You can visit the ranch in the fall even if you aren't participating in the clinics as it, too, is a wonderful time to be there. The adults-only format translates into a slower, more relaxed pace, and the weather is just perfect. Warm, sunny days and cool, crisp evenings bring on the spectacular fall foliage that is just breathtaking.

When most mountain ranches are shutting down for the long winter months, C Lazy U is gearing up for an activities-packed season. There are just as many things to do in the winter as there are in the summer. The ranch grooms 25 miles of trails, perfect for cross-country skiing through gorgeous scenery. There are Telemark trails, too, even for beginners. Try snowshoeing for a different experience. The ranch runs complimentary shuttles to Winter Park and Silver Creek ski areas if you feel the need for speed (downhill skiing). Riding horses through the snow is a unique experience and something most folks don't get to do. If you prefer, you can stay indoors and practice your horsemanship skills, in the 10,000 square foot arena. Then there are sleigh rides, sledding competitions and ice-skating. Evenings bring another slew of things to do. Night sledding and moonlit ski tours are the most popular activities. You can also just sit around the piano and listen to great music while sipping a hot cup of cocoa or your favorite toddy.

Elegant western dining awaits you at the end of your day. You might be treated to a Rack of Lamb with Apricot Sauce or Grilled Bear and Antelope Sausage and Venison. Don't be surprised to find ostrich or pheasant on the menu, either. For lighter appetites, the ranch features Grilled Salmon with Orange Dill Au Jus or Artichoke Stuffed Trout in a Basil Lemon Sauce.

These and many more tempting dishes are created under the creative direction of veteran Chef Bob Inman. Herding a staff of 8 to 15 (depending upon the season), Chef Inman oversees all meal periods at C Lazy U. Graduating with a culinary arts degree over 12 years ago, Chef Inman did an internship at The Homestead, a five star restaurant on the east coast. The 3-month internship turned into 4 years and Chef Inman says it's where he really learned how to cook. Interestingly, Chef Inman is listed in the Guinness World Book of Records for participating on a team that created the world's largest pecan pie. The pie was sold in $1.00 pieces and the proceeds were donated to the Make-A-Wish Foundation.

THE GREAT RANCH COOKBOOK

Dinner Menu

BACON-WRAPPED BAY SCALLOPS

MARINATED MUSHROOM SALAD

THREE PEPPERCORN ✪
GARLIC PRIME-RIB

Or

MARINATED GRILLED ✪
SHRIMP KEBABS

RANCH-STYLE GREEN BEANS ✪

WHITE AND WILD RICE
WITH APPLE AU JUS

C LAZY U CHEESECAKE

✪ Recipe Included

Bacon-Wrapped Bay Scallops

At first I thought he must mean sea scallops. How can you wrap those itty-bitty bay scallops? My tester, Kathy, told me it was doable. We did it and wouldn't you know? They were the cutest things and so scrumptious! Chef Inman serves these with Dean Martin's Oriental BBQ sauce, which I couldn't locate. We used my favorite Vidalia Onion Barbecue sauce. Try any sauce you like, barbecue or not.

4 SERVINGS

16 fresh bay scallops
8 slices bacon

Parcook bacon either in a microwave or in a 350° oven. You want it to cook half way to crispy. We cooked ours in a microwave for 3 minutes on high power. Cut the parcooked bacon in half lengthwise, then into 2-1/2" long strips. Dry scallops with paper towels. Wrap bacon around scallop and insert a toothpick to secure. Bake for 8 to 10 minutes. Chef Inman drops the scallops in a deep fryer at 375° for 2 to 3 minutes. If you use the deep fryer, you do not need to parcook the bacon.

Three Peppercorn and Garlic Prime Rib

Despite the peppercorn crust, this beef is not hot. It is peppery, and the crust is crunchy, which I really like. It's simple to make and is big on flavor. It doesn't need it but serve with a side of horseradish if you're so inclined.

6-8 SERVINGS

3 tablespoons each of white, green and black peppercorns
3 tablespoons minced garlic
1/4 cup olive oil
3-1/2 to 4 pounds standing rib roast

Roughly grind the three peppercorns in a spice grinder or pepper mill. (I put the peppercorns in a sealed plastic bag and roughly crushed them with my mallet.) In a small bowl, mix peppercorns, minced garlic and olive oil. Place roast on a rack in a roasting pan and rub the peppercorn mixture all over the meat. Place in oven and roast at 275° until you reach an internal temperature 135°-140°. Let stand for 10 minutes before carving. (I cooked a 3-1/2 pound roast for 2-1/2 hours and pulled it out at 135° for a perfect medium rare.)

B. Hylis

Marinated Grilled Shrimp Kebab

An incredible smoky tomato-garlic flavor dominates these delicious kebabs.
You can make these as an appetizer or as a main entrée as
Chef Inman has here. Try substituting a portabella mushroom cap
for the button mushrooms for a variation.

6 SERVINGS

Marinade:
1 tablespoon minced garlic
1/2 cup olive oil
1/4 cup tomato sauce
2 tablespoons red wine vinegar
2 tablespoons minced basil
2 tablespoons sun-dried tomato pesto*
Pepper to taste

Kebabs:
18 (26-30 count) fresh shrimp, peeled and
    de-veined
1 each green, red and yellow bell pepper,
    cut into chunks
12 button mushrooms

In a large bowl mix all the ingredients for the marinade together. Remove 1/2 cup and set aside. Add the shrimp to the remaining marinade, cover and refrigerate for 1 hour. In another large bowl or a plastic bag, add the vegetables and the reserved 1/2 cup of marinade. Cover (or seal the plastic bag) and refrigerate for 1 hour.

After 1 hour, skewer the shrimp and vegetables in this order: mushroom, yellow pepper, shrimp, red pepper, shrimp, green pepper, shrimp, mushroom. Place skewers on a hot grill and grill, occasionally turning, until shrimp is done, about 4 to 6 minutes.

*I found a great sun-dried tomato pesto at A.J.'s Fine Foods.
See "Sources" section on page 236.

THE GREAT RANCH COOKBOOK

Ranch-Style Green Beans

I'm reminded of my southern roots when I eat these delicious green beans. The combination of bacon and chicken broth imparts a comforting flavor to the beans and suddenly I'm yearning for West Texas stardust nights, and Mom's country cooking. This dish is best when the green beans are young and tender.

8 SERVINGS

2 pounds fresh green beans
3 strips bacon, cut into matchstick strips
1 small red onion, finely chopped
1 tablespoon all-purpose flour
3 cups water
2 chicken bouillon cubes or 2 teaspoons chicken base
Salt to taste
White pepper to taste

Cook bacon in a medium saucepan until the fat is rendered, about 8 to 10 minutes. Add onion and cook for 1 minute. Add flour and mix well before adding the 3 cups of water. Add the chicken bouillon cubes and bring to a boil. Reduce heat and season with salt and white pepper. (Don't add too much salt at this time, because the sauce will reduce some as it cooks and concentrate the salt flavor.) Add the green beans and simmer uncovered until the green beans are tender, 13 to 20 minutes. Taste and adjust seasonings if necessary.

Elk Creek Lodge

P.O. Box 130
Meeker, CO 81641
(970) 878-5454

Season: mid-June through mid-November

Capacity: 18

Accommodations: private, luxurious cabins on the White River and Elk Creek with spacious bedrooms, large sitting area

Activities: predominantly fly-fishing; hunting; for non-anglers, horseback riding; hiking; sightseeing

Rates: Fishing - $3,200 weekly per person; hunting - $4,000 weekly per person; includes all meals, lodging, guides

Elk Creek Lodge is renowned for its premier fly-fishing program. From the luxurious and spacious private cabins to the impressive rustic main lodge, to the fabulous feasts prepared by seasoned Chefs, Elk Creek is the epitome of first-class. With the abundance of private water, literally "flooded" with trophy-size trout, Elk Creek is a much sought-after destination for anglers in search of the classic Western America fly-fishing experience. For a really awesome adventure, sign up for the fly-out to the Green River on the lodge's private plane to fish what experts describe as "incomparable water." It's a float trip you will never forget.

There are activities for non-anglers, too, but how can you not fish when you are in fisherman's paradise, and you can see the fish rising to the surface to snatch an early hatch? Horseback ride through picturesque wilderness or hike through pine-scented forests, and wildflower-drenched meadows if the fishing bug doesn't bite you. For a special treat try a fly-out shopping jaunt to Aspen or Vail, or both!

Regardless of how many activities you choose, one thing for sure is that one of your favorites will be eating. Goldie Veitch and Charles Clement make sure you leave the lodge in awe of their combined creative talents. Breakfast and lunch are Chef Veitch's responsibility and one she takes most

seriously. She has been cooking for over 20 years and has developed quite a large fan club of guests drooling over her special pancakes and muffins and egg dishes. Over the years, so many guests asked for her morning recipes that she finally put together a fabulous cookbook containing all the favorites. You can order Good Morning, Goldie! and create the same famous breakfasts at your home. Just write to Goldie Veitch at P.O Box 1424, Meeker, CO 81641 and include a check for $11.95, which includes shipping and handling. Or you can pick up a copy at the lodge while you're there.

Evenings are relaxing at the lodge and by, oh, 5:30 or 6:00 p.m., the smells wafting from the kitchen are enough to make you lose your train of thought. Anticipating the coming meal is a sport in and of itself. What culinary masterpiece will Chef Clement create tonight? Whatever it is, you know that it will be a palate-pleasing cornucopia of gastronomic delight. The menu Chef Clement has shared with us is a fine example of the depth and breadth of his creative talents. His dishes are full of flavor and texture, and show guests that American West cuisine is just as prolific in taste dimensions as any other regional cuisine.

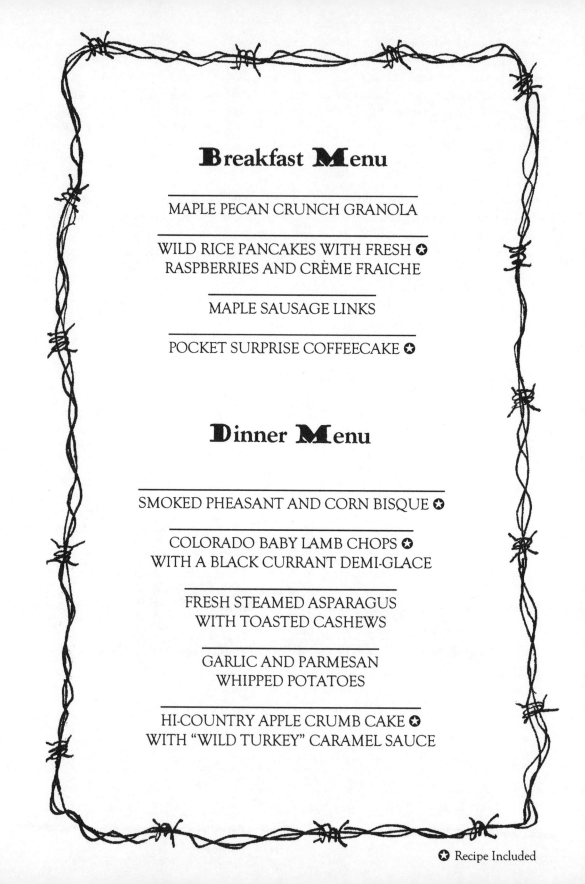

Breakfast Menu

MAPLE PECAN CRUNCH GRANOLA

WILD RICE PANCAKES WITH FRESH ✪
RASPBERRIES AND CRÈME FRAICHE

MAPLE SAUSAGE LINKS

POCKET SURPRISE COFFEECAKE ✪

Dinner Menu

SMOKED PHEASANT AND CORN BISQUE ✪

COLORADO BABY LAMB CHOPS ✪
WITH A BLACK CURRANT DEMI-GLACE

FRESH STEAMED ASPARAGUS
WITH TOASTED CASHEWS

GARLIC AND PARMESAN
WHIPPED POTATOES

HI-COUNTRY APPLE CRUMB CAKE ✪
WITH "WILD TURKEY" CARAMEL SAUCE

✪ Recipe Included

Wild Rice Pancakes

I think these are what put my neighbor Alex over the top (of the scales, that is). I covertly invited him over to try these fabulous flapjacks while his wife Rosalie was away. I didn't know he had already eaten breakfast that morning. They were so good he had a second helping. Shortly after this rendezvous, Alex and Rosalie informed me that he could no longer come over and eat. And I wasn't even near through testing recipes! I just couldn't lose my best taster. (Don't worry, turns out I didn't.)

Back to the cakes, these are so very, very good. I absolutely love the crunchy texture the rice provides and you won't believe the way they taste with pure maple syrup. You'd think wild rice and maple syrup were made for each other like peanut butter and jelly.

YIELD = 8-10 PANCAKES, 4" EACH

3 eggs at room temperature
1-1/2 cups milk, warmed
1/2 cup butter (1 stick) melted
2 cups flour
3 teaspoons baking powder
1/2 teaspoon salt
1/2 teaspoon sugar
1 cup cooked and cooled wild rice (about 1/2 cup uncooked)

In a large bowl, beat the eggs and milk together. Stir in melted butter. In a separate bowl, mix the flour, baking powder, salt and sugar together. Pour egg mixture over flour mixture and stir until smooth. Stir in cooked wild rice. The batter will be fairly thin.

Slowly ladle 1/4 cup or so onto a medium hot griddle or skillet and cook 2 to 3 minutes or until golden brown and flip. Brown on other side. Scoop batter up from the bottom, as the rice tends to fall to the bottom. Serve with butter, warm maple syrup and fresh raspberries, blueberries or strawberries.

Pocket Surprise Coffeecake

Good warm or cold, this soft, yeasty dough is tasty and smells heavenly
while baking. It's kind of sticky work, but it's fun poking holes in the dough
and filling them and then covering them up.
(Sounds like my dog burying a bone!)

YIELD = A 9" X 9" CAKE, CUT INTO 9 PIECES

1 cup warm water
1 tablespoon yeast
1/2 cup sugar

1 egg, lightly beaten
1/4 cup butter (1/2 stick) melted
1 teaspoon grated lemon peel
1/2 teaspoon salt
2-1/2 cups flour, divided

1 cup peeled, cored, chopped apple
2 tablespoons sugar
1/2 teaspoon cinnamon
2 tablespoons flour

1 cup powdered sugar
2 tablespoons milk
1/3 cup chopped pecans, toasted*

Grease a 9" X 9" square baking pan. In a large bowl, place warm water, yeast and sugar
and stir. Set aside for 10 minutes. Stir egg, butter, peel, salt and 1-1/2 cups of the flour
into the yeast mixture. Mix well. Add enough flour for a soft, but still sticky dough.
Spread into the greased pan. Cover and let rise about 1 to 2 hours, or until it rises just
over edge of pan.

Mix apples, sugar, cinnamon and 2 tablespoons flour in a small bowl. Preheat the oven
to 375°. Poke 9 holes (equally spaced apart, 3 across, 3 down) in the dough. It will fall
when you do this and that's okay. Dough will also stick to your fingers, so you can grease
them a little. Using a spoon, drop about 2 tablespoons of apple mixture into the holes
you poked. If the apple doesn't go all the way down the hole, that's okay. Flour your
fingers and gently poke the apple mixture down while pulling a bit of the dough over the
apple mixture. It will cover the apples fairly well. You'll get the hang of it after a hole or
two. Bake for 25 to 35 minutes or until golden brown. Remove from oven and cool
slightly.

Mix powdered sugar and 2 tablespoons milk until smooth. Drizzle over the top of the
cake and sprinkle with pecans.

To toast pecans, see page 11.

THE GREAT RANCH COOKBOOK

Smoked Pheasant and Corn Bisque

My friend and favorite Brew Master Sean McLin told me <u>after</u> the fact that I didn't want to use the darkest beer (Guinness Stout) for this recipe because it turns bitter as it reduces. Thanks, Sean. On to round two, I tried a lighter dark beer and it was mucho better. Done correctly (sans stout), this is a fantastic soup! The name alone sounds regal and the taste is definitely royal. Thick and slightly chunky from the pheasant, I have to rank this soup as one of the top selections in this book, and another reason why I think Chef Clement is an artist in Chef's clothing. You can smoke your own pheasant or buy one already smoked. A.J.'s Fine Foods carries both versions if you can't find a source in your area. See the "Sources" section on page 236 for details.

YIELD = 1-1/2 QUARTS, 6 8-OUNCE SERVINGS

1 pheasant, smoked and cut into quarters (discard back piece)
2 cups dark beer (amber ales work best)
3 ears fresh corn
1/2 stalk celery, chopped
1/2 carrot, peeled and chopped

1/2 medium onion, chopped
1 medium potato, peeled and chopped

2 cups heavy cream
Salt and pepper to taste
2 tablespoons chopped parsley (optional)

Simmer pheasant in beer approximately 40 minutes. Remove pheasant to cool and continue to simmer beer until reduced by 1/2, or about 1 cup is left. Remove skin from pheasant and chop meat. Set aside. Cut corn kernels from cobs and add corn kernels, celery, carrot, onion and potatoes to a medium stock pan. Add reduced beer and just enough water to cover the vegetables. Bring to a boil, reduce heat and simmer until vegetables are very tender. Add cream and stir. In batches, transfer soup to a blender and puree until very smooth. Return soup to stock pan and add chopped pheasant. Return to a simmer and season with salt and pepper. Garnish with a sprinkling of chopped parsley.

Colorado Baby Lamb Chops

Generally, I don't like food with the word "Baby" in the title. But, I have to open that narrow little mind of mine because these chops are "ooh baby, good!" My dear friends Bonnie and Art Cikens volunteered to taste these morsels because they love lamb and can recognize a good piece of "bah bah black sheep" when they see it. They were drooling over this dish. Bonnie kept saying it was the best meal she'd had since she'd been in Scottsdale. The Black Currant Demi-Glace is a complex, ravishing sauce and it makes the lamb chops shine.

4-6 SERVINGS

4 lamb racks, 4 ribs each
3 cloves garlic, minced
1 teaspoon fresh thyme
1 teaspoon fresh rosemary, chopped
Salt and pepper to taste

Sauce:
1 cup dried black currants
1 teaspoon olive oil
1-1/2 teaspoons minced shallots
1-1/2 teaspoons minced garlic
1/2 cup red wine
2 cups Demi-glace*
1 cup dried black currants
1/2 teaspoon fresh thyme

Preheat oven to 350°. Mix garlic, 1 teaspoon of fresh thyme, and rosemary together. Rub mixture on lamb racks and season with salt and pepper. Heat a sauté pan over medium high heat until very hot. Add racks and brown quickly on all sides. Transfer racks to oven to finish cooking (check temperature after 10 to 15 minutes). Remove from oven when temperature reaches 125°-130° and let rest 5 to 8 minutes before carving into individual chops.

As chops go into the oven, start sauce by heating olive oil in a medium sauce pan. Add shallots and cook 1 minute. Add garlic and cook another minute. Add red wine and bring to a boil. Cook until wine is reduced to 2 tablespoons. Add demi-glace, currants, 1/2 teaspoon fresh thyme and bring to a boil. Reduce heat and simmer until thick and syrupy.

Demi-glace is available from More than Gourmet. See "Sources" section on page 236 for details. You will need 2 (1.5-ounce) tins of Demi-Glace Gold® for this recipe.

Hi-Country Apple Crumb Cake with "Wild Turkey" Caramel Sauce

Awesome! Frankly, everything Chef Clement submitted to this book is spectacular, so I'm not surprised his dessert is prodigious. The cake is super-moist, filled with chunky pink-tinged apples and sauced with buttery, liquor-infused caramel. This dessert won kudos from Pam Raby, the assistant director at my health-club, not because it's healthful, but because it's not! It just tastes great. Of course, she can afford to eat all she wants. She's not got an ounce of body fat on her. Have some more cake, Pam.

12 SERVINGS

2 cups sugar
1 cup vegetable oil
3 eggs
3 cups all-purpose flour
1 teaspoon baking soda
1 teaspoon salt
1-1/2 pounds Rome apples (about 3 to 5 apples), cored and chopped but unpeeled
1-1/2 tablespoons lemon juice

Topping:
1/2 cup butter (1 stick)
1 cup sugar
3/4 cup flour
1 tablespoon cinnamon (optional)

Beat sugar, oil and eggs together for 2 minutes. In a separate bowl, mix flour, baking soda and salt together. Add flour mixture, apples and lemon juice to sugar mixture (batter will be very stiff). Pour into a greased 9" X 13" baking dish.

Mix topping ingredients together until it resembles coarse meal. Sprinkle evenly over top of cake batter. Bake 40 to 55 minutes, or until golden brown on top and a toothpick inserted in the center comes out clean. Serve with "Wild Turkey Caramel Sauce" (recipe follows).

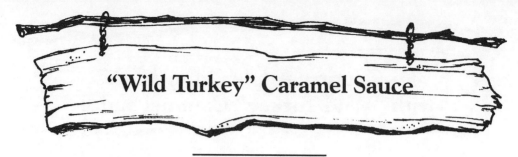

"Wild Turkey" Caramel Sauce

YIELD = 1-1/2 CUPS

1 cup sugar
2 tablespoons water
1 cup heavy cream, slightly warmed
1/4 cup butter (1/2 stick)
2 tablespoons Wild Turkey or other bourbon

Cook sugar and water over medium high heat. (It's okay if the sugar crystallizes. Continue cooking and eventually it will re-melt and turn golden brown). When sugar turns golden brown, lower heat and slowly and carefully add cream, stirring constantly. Watch out for a violent reaction between the hot sugar and the cream. Continue cooking to dissolve any sugar that seized up during the cream addition. Add the bourbon and remove from heat. Drizzle over Hi-Country Apple Crumb cake and garnish with a slice of fresh apple.

B. Hillis

Elk Mountain Ranch

P.O. Box 910
Buena Vista, CO 81211
(719) 539-4403
(800) 432-8812
Website: http://
　　www.elkmtn.com
E-mail: elkmtn@sni.net

Season: End of May through
September

Capacity: 30

Accommodations: Log cottages
and lodge rooms, carpeted, and
tastefully furnished with daily
housekeeping. Cottages are large
enough to accommodate family
groups of 4-6 persons.

Activities: Horseback riding
through unspoiled wilderness;
thrilling whitewater rafting;
excursions to Aspen, CO;
overnight camping trip high in
the Rocky Mountains; extensive
children's program; two stocked
fishing ponds; trapshooting;
archery; hiking; square dancing;
volleyball; Jacuzzi and more!

Rates: Weekly (seven nights),
$935-$1,100 for adults and
$655-$855 for children

ELK MOUNTAIN RANCH

Elk Mountain Ranch has the distinction of being
the highest guest ranch in Colorado, at 9,535 feet
above sea level. Secluded in the midst of the
beautiful San Isabel National Forest, the ranch is
surrounded by sweeping views and snow-capped
mountains. Nestled among aspen and evergreen
on Little Bull Creek, Elk Mountain Ranch is
paradise for photographers and nature lovers.
Deer, elk, antelope and wildflowers abound
throughout the property. Around the turn of the
century the ranch was once the mill site for the
Futurity Mining Camp. The Trading Post (ranch
store) is an original prospector's cabin. The ranch
takes only a maximum of 30 guests at a time, and
by the end of the week, everyone feels like they are
part of one big family. Sue and Tom Murphy are
natural-born people-persons and they have a knack
for finding staff that shares the same philosophy as
they do – the guests' satisfaction is priority one.
Finding fresh flowers and fruit in your room upon
check-in sets the stage for a week of pampering
and attention.

With so many activities to choose from, you'll
surely be back for a second visit. The ranch will
schedule the week's activities for you or you can
just leisurely do nothing. The riding program is
outstanding, with both on and off-trail riding.
Being able to wander off the trail and explore the
glory of the unspoiled forest is a treat not many
ranches offer. You can even take an instructional
class on the tack, care and feeding of horses. There
is an overnight camping trip that begins with a
pristine ride to a secluded campsite for an evening
of star-gazing, story-telling and a cookout that will
make you never want to eat again. Well, at least
until you smell the bacon and eggs at the next
morning's breakfast cookout!

If you want to explore activities other than riding, Sue says not to miss the whitewater-rafting trip. Adventure awaits you on the rapids of the roaring Arkansas River through some unbelievable canyon scenery. If the rapids don't take your breath away, the views certainly will. The area is rich in Ute Indian and mining history. Special hikes are geared around historic Indian battlegrounds, where prized arrowheads can still be found. Great fishing streams are located nearby, or you can fish the ranch's two stocked ponds. If you'd like, the ranch will teach you how to clean your trout and take them to the kitchen to cook them for you. The ranch also provides instruction and equipment for archery and trapshooting. I can't imagine you would want to leave the ranch for anything, but if you do, there are excursions to Aspen and Breckenridge for sightseeing and shopping.

All the activities will work up a mighty appetite and the ranch's two cooks have just the remedy for you. The cuisine is straightforward, delicious and filling. Homemade soups, fresh baked breads and juicy steaks and roast turkey await you at the end of the day. Save room for the special homemade desserts. Saturday evening, a candlelight French Country dinner is served to crown the week. One guest from Indiana wrote, "There's no better way to learn about a region than to meet its people and sample its cuisine. Your ranch is heaven on Earth!" Sue shares two menus with us. The nice thing about most of these recipes is that you can do much of the preparation early, saving time for you to enjoy your company instead of cooking.

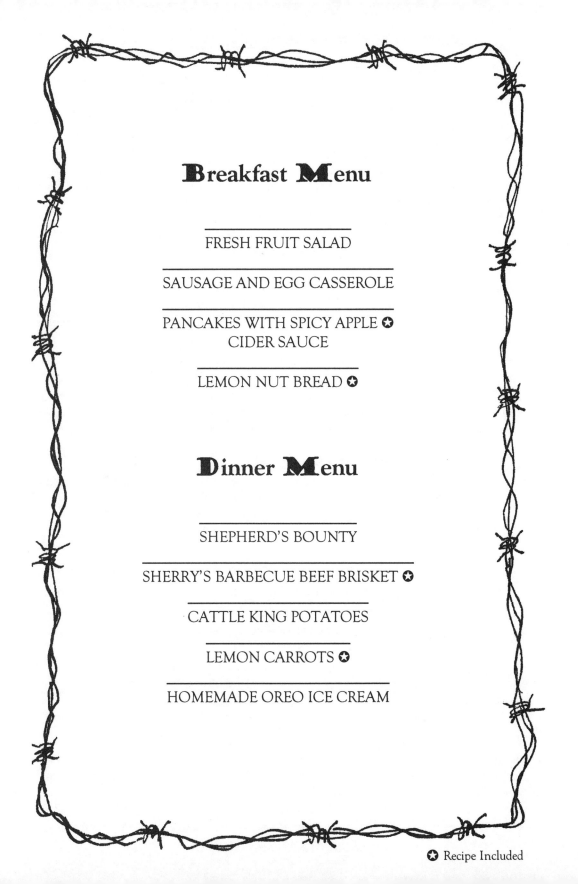

Breakfast Menu

FRESH FRUIT SALAD

SAUSAGE AND EGG CASSEROLE

PANCAKES WITH SPICY APPLE ✪
CIDER SAUCE

LEMON NUT BREAD ✪

Dinner Menu

SHEPHERD'S BOUNTY

SHERRY'S BARBECUE BEEF BRISKET ✪

CATTLE KING POTATOES

LEMON CARROTS ✪

HOMEMADE OREO ICE CREAM

✪ Recipe Included

Spicy Apple Cider Sauce

The smell of this sauce will remind you of crisp fall mornings and warm Indian summer afternoons. Pour this thick, aromatic sauce over your favorite pancakes. I even used this sauce as a dessert sauce for Wit's End's Apple Dumpling dessert. If you're in the mood, toss in 1/2 cup of toasted pecans for another flavor dimension.

YIELD = 3 CUPS

1 cup sugar
3 tablespoons biscuit mix
1/4 teaspoon cinnamon
1/4 teaspoon nutmeg

2 cups apple cider
2 tablespoons lemon juice
1/4 cup margarine or butter (1/2 stick)

Mix sugar, biscuit mix, cinnamon and nutmeg in saucepan. Stir in cider and lemon juice. Heat over medium-high heat, stirring constantly, until mixture thickens and boils. Boil 1 minute longer. Remove from heat and stir in margarine or butter. Return to heat and simmer until desired consistency is reached. (You may not have to simmer anymore if you like the sauce's consistency after you've added the butter).

Lemon Carrots

The lemon and sugar blend well together with the natural sweetness of baby carrots for a surprisingly effervescent taste. Garnish these carrots with some chopped parsley and a lemon wheel.

4 SERVINGS

2 cups baby carrots, washed
1/2 teaspoon salt
2 tablespoons sugar

2-1/2 tablespoons fresh lemon juice
4 tablespoons butter

Boil carrots in water salted with 1/2 teaspoon of salt until crisp-tender, about 3 to 4 minutes. Drain. In a saucepan, combine sugar, lemon juice and butter and heat until boiling. Boil for 2 to 3 minutes or until thickened slightly. Stir in carrots and cook until heated through, about 2 to 3 minutes.

Lemon Nut Bread

Zippy, zingy and full of pucker power! This bread has a refreshing lemony flavor that's equally at home at brunch or breakfast. I love the addition of nuts, as most lemon breads I've seen do not have nuts. A tablespoon of poppy seeds would be a nice addition as well. The key to great quick breads is not to overmix. Mix until the ingredients are just barely blended.

YIELD = 2 LOAVES

1/2 cup margarine or butter (1 stick), softened
1-1/2 cups sugar
3 eggs
2-1/4 cups all-purpose flour
1/4 teaspoon salt
1/2 teaspoon baking soda
1/2 cup buttermilk
Grated rind of 1 lemon
3/4 cup chopped pecans or walnuts (optional)
Juice of 2 lemons
3/4 cup powdered sugar

Preheat oven to 325°. Cream together margarine and sugar. Add the eggs one at a time and blend well. In a separate bowl, combine flour, salt and baking soda. Alternately add buttermilk and dry ingredients to the creamed mixture, beginning and ending with the buttermilk. Stir in grated rind of lemon. Fold in pecans or walnuts.

Spoon batter into greased and floured loaf pans, 1/3 full (this will make for a relatively flat bread, but it works well for this bread). Bake 50 to 60 minutes or until a toothpick inserted in the center comes out clean. Cool 15 minutes and remove from pan. Stir the juice of 2 lemons and powdered sugar together in a small bowl. Punch lots of holes in the top of the warm bread with a toothpick and pour on glaze. (Make sure you have a sheet pan under the cooling rack to catch the glaze).

Sherry's Barbecue Brisket

Good brisket recipes are hard to find. Here's one that will swear you off all others. With the long, slow cooking time, the meat is so tender it shreds with a fork. The sauce is wonderfully tangy and slightly sweet. Brisket is great as leftovers, so if your crowd is fewer than 10, go ahead and make this recipe, as is, for a terrific lunch tomorrow. By the way, Sherry used to be a guest of Elk Mountain; now she's a neighbor!

10 SERVINGS

5 to 6 pound beef brisket
1/4 cup water
1 large onion, peeled and sliced
1 (8-ounce) bottle chili sauce
4 cloves garlic, minced
2 bay leaves

1/2 cup brown sugar
1/3 cup Dijon-style mustard
1/4 cup red wine vinegar
3 tablespoons molasses
1/4 cup soy sauce
1/2 bunch parsley sprigs

Preheat oven to 325° with the rack in the lower third of the oven, but not at the very bottom. Sear the meat, fat side down first, in the bottom of a very hot heavy-duty roasting pan. Turn meat over and sear the other side. This searing can also be done on a grill if you prefer. Stir together the remaining ingredients and pour over brisket. Cover and cook for 4 or 5 hours, or until tender.

Remove the meat from the pot and pour the sauce into a bowl. Discard the bay leaves and cool sauce. Slice the meat when cool. Skim the fat off the sauce and pour sauce over sliced meat. Reheat meat on the stove or in the oven, covered. Garnish with parsley sprigs.

King Mountain Ranch

P.O. Box 497
Granby, CO 80446
(970) 887-2511
(800) 476-5464
Website: http://
    www.kingranchresort.com
E-mail:
    hosts@kingranchresort.com

Season:
Mid-May through October

Capacity: 65

Accommodations: 2 lodges with 34 rooms. Rooms either have a king-size or two double-size beds and all have private baths. Every room has a view of the valley and surrounding mountains. Family units have adjoining doors.

Activities: Skill-level based riding, fly-fishing, hiking, swimming, tennis, recreation center with bowling, pool table, ping pong table and board games, library, skeet and trap shooting, golfing, whitewater rafting, hot air ballooning. Conference facilities are available.

Rates: Mid-June through mid-September - $1,200 per adult, double occupancy 7-nights; children rates $500 or $675, depending upon age; call ranch for 3-night packages and discounted off-season rates; special rates for Au Pair/Nanny.

KING MOUNTAIN RANCH

Located in northwest Colorado, near Rocky Mountain National Park, the King Mountain Ranch provides families and corporate groups with a spectacular setting in a lush, pine-covered valley. Only 90 minutes from Denver, the ranch is close to civilization yet feels thousands of miles away. The entrance to the ranch is four miles from the highway and those four miles begin to melt away the outside world with every turn, from the first view of the private lake to grand lodges tucked into the mountainside among tall pines and shimmering aspens.

At 9,000 feet above sea level, this mountain setting is perfect for a variety of activities for your family. The ranch offers an extensive riding program, matching your skill level to the horse you'll have for the week and the type of riding you'll do. Be sure to make the lunch ride to Inspiration Point for views that will bring a new perspective to what life is all about. Children get special attention at the Foxes Den, an activity center just for them. Enjoy fishing for rainbow, cutthroat and brook trout on the 50-acre stocked lake. Stream fishing on Willow Creek is also available for the fly-fishing enthusiast. If fishing and riding aren't enough, there also is hiking on numerous trails on the ranch and through Arapaho National Forest. An indoor heated swimming pool and Jacuzzi help nurse the sore muscles that come with a full day of activity. Two lighted tennis courts and an antique bowling alley add other diversions to immerse yourself in while at the ranch. Off the ranch, but close by, are whitewater-rafting and golfing on Colorado's number one public course. If shopping is part of your normal vacation activities, there is the King Mountain Ranch store located next to the office and you'll find contemporary western jewelry, clothes, and unique gift items to bring back to your friends and family.

The King Ranch hosts several family reunions each year, and with groups of 30 or more you can enjoy exclusive use of the ranch. For business meetings, the ranch has several conference rooms, audio-visual equipment and an attentive staff that will make you think about opening a branch office right at the ranch!

The staff is a key element to the success of any ranch and King Mountain interviews over 500 college men and women, selecting the best 30 or 40 each year. In addition to selecting the staff based on warm personalities and flexibility, the ranch also takes into consideration the applicants' commitment to their own communities, college grade point averages and references. Once selected, the staff is empowered to make decisions that will make each guest's stay memorable. Most of the guests' comments acknowledge the outstanding staff at King Mountain Ranch. One recent guest from Massachusetts wrote, "The Ranch: wonderfully run, amazing children's program, super staff, fantastic food, the best!!"

Speaking of food, the ranch's Executive Chef Mark DeNittis spares no expense in combining the freshest local ingredients with classical techniques to provide an experience worth repeating. A 1992 graduate of Johnson & Wales, Chef DeNittis apprenticed in Italy and honed his culinary skills at the famous Breakers Hotel Resort in Palm Beach. Chef DeNittis suggests that you relax in the Trail's End bar, unwinding by the fire before adjourning to the dining room with its large picture window capturing majestic views of the valley and surrounding mountains. The cuisine has a definite southwest influence, but other cuisines are fused and blended in to create unique dishes that are signature to the ranch. You'll find wild game tamed in such dishes as Blackened Buffalo Quesadillas and Grilled Raspberry Wheat Beer Quail. Classic favorites are also served with a special "King Ranch" flare, like Trout a la Meniere and Stuffed Chicken Breast. Emphasis is on fresh ingredients, combined with creative interpretation and artful presentation. Interspersed between the gourmet meals at the lodge are weekly traditional western-style barbecues served outdoors under the stars and captivatingly authentic.

Breakfast Menu

COWBOY BISCUITS AND GRAVY
Or

SMOKED SALMON AND ONION ✪
OMELET WITH SOUTHWEST
HOLLANDAISE
Or

HUEVOS RANCHEROS

Dinner Menu

BLACKENED BUFFALO ✪
QUESADILLA WITH GOAT CHEESE

GRILLED 10-OUNCE RIB-EYE ✪
WITH TOBACCO ONIONS

GRILLED SUMMER VEGETABLES

MESA VERDE RICE ✪

ALMOND CHEESECAKE ✪

✪ Recipe Included

Smoked Salmon and Onion Omelet

with Southwestern Hollandaise

Having someone special for breakfast or brunch? This is the perfect "I think I'll impress you" dish. The colors and presentation are breathtaking and the smoky flavor is haunting. If you don't want to execute the hollandaise the old-fashioned way, just buy (forgive me Chef Rosenberg) a pre-packaged plain hollandaise mix and add the spices and orange juice.

3 SERVINGS

3 - 6 slices smoked salmon (about 4 ounces)
1/2 tablespoon butter
1/4 cup chopped spinach
1/4 cup chopped mushrooms
2 tablespoons finely chopped red onions
2 tablespoons cream cheese, very soft
2 to 3 eggs per serving (6 to 9 total)
1/4 cup Southwestern Hollandaise per serving (3/4 cup total)

Cook onions in butter until they are translucent, about 3 minutes, then add the spinach and mushrooms. Simmer until the mushrooms are tender, 3 to 4 minutes. Prepare an omelet by pouring 2 to 3 beaten eggs into an 8" non-stick skillet over medium heat. As the omelet begins to cook, run a spatula around the edges, allowing the liquid eggs to seep to the edges and underneath the edges. As the edges begin to cook and very little liquid eggs are left, flip the omelet over. Spread cream cheese until it covers 1/2 of the omelet and add salmon on top of cream cheese then mushrooms, onions and spinach mixture. Place in 350° oven for 3 minutes to make sure the eggs are cooked through. Remove from the oven and fold in half onto plate. Ladle the Southwestern Hollandaise (recipe follows) on top and serve.

Southwestern Hollandaise

Achiote is used in Mexican and Mayan Indian dishes, and is the key ingredient in "Adobado" style sauces. The color of this paste is dark brick red and usually comes in small, thin wrapped packages. AJ's Fine Foods carries the Achiote Condimentado brand and will ship it to you. See more details about AJ's in the "Sources" section on page 236.

YIELD = 1 CUP

4 egg yolks
1/4 teaspoon salt
1/8 teaspoon cayenne pepper
2 teaspoons diluted achiote condimentado (annatto seed paste*)
1 cup hot butter, divided
2 tablespoons fresh lemon juice
2 tablespoons orange juice
2 tablespoons chopped cilantro plus extra for garnish

Beat egg yolks with a whisk until thick and lemon-colored. Add salt, cayenne pepper and achiote. Add 1/2 cup butter, 1 tablespoon at a time, whisking constantly. Combine the remaining butter with lemon and orange juice. Slowly add to mixture 1 tablespoon at a time, whisking constantly. Stir in cilantro. Ladle over omelet and sprinkle a little more chopped cilantro on top.

*Achiote paste is diluted with equal amounts of water. For 2 teaspoons of diluted achiote paste, mix 1 teaspoon of paste with 1 teaspoon of water. Achiote paste is available through A.J.'s Fine Foods. See "Sources" section on page 236 for details.

Blackened Buffalo Quesadillas

A western twist on a Mexican favorite. With its low fat and cholesterol, excellent flavor, and old west appeal, buffalo meat works great for King Mountain Ranch's "gourmet western cuisine." I ordered buffalo through AJ's Fine Foods. At the time I couldn't get a tenderloin, so I opted for a New York strip cut of buffalo and it worked perfectly in this recipe. If you can't get buffalo at all, beef is equally good. We experimented with different cheeses and prefer Boursin, Feta or herbed cream cheese to goat cheese. Add other condiments as well if you like, such as guacamole, salsa, caramelized onions, etc.

---

4 SERVINGS

1-1/2 pounds buffalo medallion or 2 (8 or 9-ounce NY strip steaks)
2 tablespoons blackening seasoning (more or less, depending upon your taste)
4 ounces goat cheese
4 each flour tortillas

Preheat cast-iron skillet to medium high. Season buffalo with blackening seasoning and cook to desired temperature. Buffalo tastes best at medium rare. For a steak, cook 3 or 4 minutes on each side and finish in a 375° oven for 2 to 5 minutes. For a tenderloin, cook about 20 minutes in total, turning every 5 minutes. Remove when internal temperature reaches 130° for medium-rare. Cut into slices and reserve. Spread goat cheese on the flour tortilla. Place buffalo slices across half of the tortilla and fold in half. Place tortilla on grill or in a skillet and heat to desired crispiness.

Grilled Rib-Eye
with Tobacco Onions

This may be the best rib-eye steak I have ever had. This is a major statement because I'm a serious steak connoisseur from way back. The liquid smoke in the marinade makes the steak taste like it was just grilled over crackling mesquite under the star-splattered midnight blue sky. I can hear the serenading cowboy now. The tobacco-colored onions are a fantastic addition to this authentic Western treasure.

6 SERVINGS

Steak Marinade:
1 cup liquid smoke (preferably mesquite-
   flavored)
1 cup Worcestershire sauce
1 cup orange juice
1/2 cup brown sugar

1 cup soy sauce
1/4 cup vegetable oil
1 tablespoon jerk seasoning*
2 tablespoons cracked black pepper
1 shallot, minced
1 clove garlic, finely chopped

Combine all ingredients and whisk until blended.

6 (10-ounce) rib-eye steaks

Tobacco Onions:
3 onions, sliced paper thin
1 cup all-purpose flour (or more if needed)
1 teaspoon salt
3/4 teaspoon black pepper

Marinade steak for at least 24 hours. Preheat grill to medium high. Grill until desired temperature is reached (for a 3/4" steak, about 4 to 5 minutes on each side for medium-rare). Toss onion rings in flour until lightly covered. Deep fry in 350° oil until brown. As soon as you remove onions from oil, season with salt and pepper. Sprinkle onions over steak and serve.

Jerk seasoning is a blend of several herbs and spices, including ginger, brown mustard, allspice, fennel, thyme, red and black pepper and cloves. It's used extensively in Caribbean cooking. I found a great blend at A. J.'s Fine Foods. See "Sources" section on page 236 for details.

Mesa Verde Rice

I love this twist on Spanish rice. Rich tomato flavor with a nice spice kick,
it's perfect with the grilled rib-eye entrée. The rice will be slightly sticky,
perfect for molding for dramatic presentation. Just scoop the completed rice
into a small ramekin and turn upside down on your plate. Lift off the
ramekin and you have a perfectly shaped mound of rice.

8 SERVINGS

2 roma tomatoes, cored and quartered
1 clove garlic
3 sprigs of cilantro
1/4 cup chopped red onion
1/2 jalapeno, seeded
1/2 teaspoon salt

1/4 teaspoon pepper
2 teaspoons butter
1 teaspoon vegetable oil
2 cups white rice
2-3/4 cups chicken stock*
3/4 cup tomato juice

Puree tomatoes, garlic, cilantro, red onion, jalapeno, salt and pepper in a food processor
or blender. In a saucepan, heat the butter and oil and add the rice. Cook, stirring
frequently about 2 minutes. Add pureed ingredients and cook another 2 minutes,
stirring. Add chicken stock and tomato juice. Bring to a boil, stirring constantly. Turn
down the heat, cover and let simmer for 25 to 30 minutes until liquid is absorbed. The
rice will be a little sticky but it shouldn't feel gummy, a sign of overcooking.

*Concentrated chicken stock is available through More Than Gourmet. See
"Sources" section on page 236 for details. One (1.0-ounce) tin of Fond de Poulet
Gold® makes 5 cups of chicken stock.*

Almond Cheesecake

The presentation of this cake is so lovely. Everyone I gave it to said, "Wow, that's beautiful." I got the same comments after they tasted it. The cheesecake is a good basic cheesecake recipe, but when you add the brandy-laced nut sauce and peach slices, the cake rises to another level.

YIELD = 1 CHEESECAKE

For crust:
1/4 cup sugar
1-1/4 cups graham cracker crumbs
1/4 cup butter (1/2 stick) or less, melted

For cheesecake:
2 pounds cream cheese, at room
    temperature
1 cup + 1 tablespoon sugar
1 tablespoon cornstarch
2 tablespoons heavy cream
3 eggs
1/4 teaspoon almond extract

For nut melange:
3/4 cup sugar
1/2 cup water
1/4 cup chopped walnuts
1/4 cup chopped pecans
1/4 cup slivered almonds
2 tablespoons sun-dried cherries
2 tablespoons brandy
1 can peach halves, drained

For crust: Preheat oven to 350°. Mix sugar and graham cracker crumbs together. Add just enough melted butter to moisten. Press crust into bottom of a 10" spring-form pan. Bake 5 minutes. Remove from oven to cool and reduce oven heat to 275°.

For cheesecake: cream sugar and cream cheese together in a mixer, scraping sides. Mix in cornstarch and cream. Slowly add eggs while mixing, stopping to scrape sides. Pour mixture into prepared cake pan. Place in oven. On a rack just below the cheesecake, place a large roasting pan filled with hot water (to create moisture in the oven while the cheesecake is baking). Bake in a 275° oven for 1 hour and 45 minutes. Remove from oven, cool completely then chill.

For nut melange: Bring sugar and 1/2 cup of water to a boil. Boil until slightly thickened. Mix 1/2 cup of this sugar syrup with remaining melange ingredients (except peach halves) and set aside.

Remove cheesecake from pan. Garnish with melange and peach halves.

Latigo Guest Ranch

P.O. Box 237
Kremmling, CO 80459
(970) 724-9008
Website: http://www.dude-
ranch.com/latigo ranch.html
E-mail: latigo@compuserve.com

Season: June-September as a
Guest Ranch; December 15-
April 1 as a Nordic Center

Capacity: 40

Accommodations: 3-bedroom
cabins (6); 1-bedroom
accommodations (4)

Activities: Winter — cross-country
skiing; snowshoeing; tubing;
Summer — horseback riding;
trout fishing; hiking; overnight
pack trips; rafting; and children's
program

Rates: Adults — $1,495/person
per week; Children — varies
according to age. (Discount rates
available for certain dates, call
ranch for details

Seconds after you enter the Latigo gate you are gazing
out at a 200-mile panoramic view of the spectacular
Rocky Mountains. You're dazzled that you just
moments ago turned off a civilized paved road onto a
dusty trail called Red Dirt Road. Climbing up and
out of sagebrush and into alpine evergreens, the path
to Latigo is a mercurial scene, winding its way toward
quite possibly the most tranquil environment you've
ever experienced. You know instantly that Latigo is a
special place, where at 9,000 feet people want to
linger forever soaking up the breathtakingly beautiful
views of lush aspen groves, wide expanses of
wildflowers, remote beaver ponds and high vistas.
The clean, crisp air and the soul-soothing serenity of
a setting surrounded by national forests provide
guests with an unforgettable feeling of peace and
intimacy. The warm and friendly staff and owners
strive constantly to assure that this remarkably
unique ranch offers something for everyone, winter
or summer.

Once at the ranch, many guests completely absorb
themselves in the spectacular riding program. There
are sunrise rides, sunset rides, overnight rides, family
rides, loping rides, breakfast rides, dinner rides,
romantic rides. You name it and Latigo has a ride
perfect for any occasion. If you want to do some
bona fide cattle roundup riding, well, the Latigo's got
that, too. The horses are paired with the individual
riders for the entire stay based on skill and
personality (of both the rider and the horse!). If you
do decide to leave the ranch to explore the
surrounding territory, less than a couple of hours
away are Vail and Beaver Creek, Steamboat Springs
and Silverthorne. The Rocky Mountain National
Park is an easy drive as well.

THE GREAT RANCH COOKBOOK

Besides being renowned for its riding program, the Latigo is also treasured for its children's program. Many guests who travel to the ranch without children barely notice that kids are present. It's a wonderful program for both the children and the parents, both coming away with precious moments shared together and separately.

Hosts Lisa and Randy George describe their food as "both healthful and appealing." Past guests aren't quite as modest about describing the food, as evidenced by a comment from a Texas couple that while the horses were fabulous and the accommodations were warm and cozy, it was the food that was stellar. In an excerpt from a letter to friends this Texas couple was singing Latigo's praises. "I saved the food for last because it truly deserves very special attention — it is genuine gourmet — not just in name but in reality. What else can I say? I wish I owned stock in the place." College students are trained to staff the kitchen, though one of the owners is usually in the kitchen as well. Breakfast consists of either cereal and a baked goods buffet, or orders from a full-service grill. Lunch features a bountiful salad and sandwich bar plus a hot dish (such as lasagna), but Tuesdays are reserved for a cookout featuring buffalo burgers. For dinner, a choice of entrees is provided so that guests avoiding red meat may have a delicious alternative. One of the most popular dinners is the turkey dinner complete with stuffing, mashed potatoes and cranberry sauce. Lisa added, "We continually search for new recipes to add to our menu, but the most popular items remain to satisfy our returning guests' memories." I am sure that you, too, will find a Latigo vacation a memorable experience that will last a lifetime.

B.Hillis

Breakfast **M**enu

CRUNCHY GRANOLA

BROCCOLI OVEN OMELET

ORANGE GLAZED BREAKFAST
BRAID BREAD

PUMPKIN SWEET ROLLS ✪

Dinner **M**enu

MANDARIN SALAD ✪

LATIGO SUNDAY CHICKEN ✪

ALMOND BROCCOLI BAKE

HONEY GRANOLA BREAD

PEANUT BUTTER PIE

✪ Recipe Included

Pumpkin Sweet Rolls

Divine! Don't let this out, but after I tasted these, I didn't share them with anyone. I think I could eat the whole batch by myself in one sitting. The pumpkin adds a richness that's kind of earthy tasting. These freeze extremely well and are pretty enough for company. Lisa says sweet potato leftovers are a good substitution for the pumpkin.

YIELD = 20 TO 30 ROLLS

1-1/2 cups all-purpose flour
1/2 cup sugar
2 teaspoons grated lemon peel
1 tablespoon yeast
1-1/4 cups milk

1 cup pumpkin puree
1-1/2 teaspoons salt
1/2 cup margarine or oil
2-1/2 to 3 cups additional flour

Mix flour, sugar, lemon peel and yeast in a bowl. Heat the milk, pumpkin, salt and margarine to 110°, then gradually add into dry ingredients, mixing for 3 minutes on medium speed. Add additional flour to make dough elastic. Cover with plastic wrap and a towel and let rise, approximately 1 hour.

Roll out dough into a 20" X 15" rectangle. Spoon 2-1/2 cups of the topping evenly over dough. Roll up tightly along 20" side. Cut into slices, placing cut side down in pan. Cover with plastic wrap (can be refrigerated overnight) and let rise, approximately 1 hour. Sprinkle with remaining topping. Bake until brown, approximately 25 minutes. Drizzle with glaze.

Crumb topping:
1-1/2 cups all-purpose flour
1 cup brown sugar
3/4 cup margarine, cut in
1/2 cup chopped nuts (walnuts, pecans or
    sliced almonds)

Mix all ingredients together until it looks course and crumbly. Using a food processor is quick and easy. Just combine the ingredients in the bowl with the metal blade and pulse 15 to 20 times.

Glaze:
1 cup powdered sugar
1/2 teaspoon vanilla
1-1/2 tablespoons milk

Stir powdered sugar, vanilla and milk together in a small bowl. Set aside, but stir again just prior to using.

Mandarin Salad

Tart and sweet at the same time, this salad is good for exercising all your taste buds. It can also be a delightful color display on your table. I prefer using a mixture of romaine and spinach only, leaving out the regular (iceberg) lettuce.

4-6 SERVINGS

Dressing:
1/4 cup vegetable oil
2 tablespoons sugar
2 tablespoons white vinegar
1 tablespoon snipped parsley
1/4 teaspoon salt
Dash black pepper
Dash red pepper sauce

1/4 cup cashews
1 tablespoon + 1 teaspoon sugar

Salad:
1 cup chopped celery with leaves
2 green onions, thinly sliced
1/4 head iceberg lettuce, torn
1/4 bunch romaine lettuce or spinach
1 (11-ounce) can mandarin oranges,
    drained

Combine the ingredients for the dressing in the blender. Mix thoroughly and chill.

Cook cashews and sugar over medium heat in a heavy saucepan, stirring constantly until sugar is melted and nuts are lightly browned, 5 to 8 minutes. Cool and break apart.

Combine celery, onions and lettuce. Arrange on salad plates and place 4 to 6 orange sections on top and sprinkle with caramelized cashews. Drizzle 1-1/2 tablespoons of dressing on each salad.

Latigo Sunday Chicken

The name may make you think this dish is a "down-home" country version of chicken. In reality, this dish is a fascinating, unique approach and Colorado's answer to Chicken Cordon Bleu. I love the taste sensations that dance in my mouth when I taste the first and last bite of this chicken.

5 SERVINGS

5 chicken breasts, boned and skinned
5 thin slices of ham
5 thin slices of Swiss cheese
1 egg white beaten with 1/4 cup of water
3/4 cup seasoned breadcrumbs

Sauce:
1/4 cup melted margarine or butter
   (1/2 stick)
1/8 cup soy sauce
1/4 cup chicken stock*
1/4 cup sesame seeds

Preheat oven to 350°. Pound the chicken breasts so they are an even 1/4" thickness. Place a slice of ham and cheese on top and roll it up, securing it with a toothpick. Dip in egg mixture then roll in the breadcrumbs, and place on a baking sheet covered with foil (to make cleaning easier). Repeat with all pieces. Before baking, pour sauce on all pieces and sprinkle with sesame seeds. Bake for 1 hour.

Concentrated chicken stock is available through More Than Gourmet. See "Sources" section on page 236 for details. One (1.0-ounce) tin of Fond de Poulet Gold® makes 5 cups of chicken stock.

The Marvine Ranch

P.O. Box 130
Meeker, CO 81641
(970) 878-5454

Season: June through October;
Special winter program; call
ranch for details

Capacity: Intact parties
of 8-10 only

Accommodations: Guest cabins
with spacious bedrooms and
scenic view of the valley

Activities: Fly-fishing; equestrian
program; hiking; mountain
biking; wildlife viewing; winter
sports of skiing, snowmobiling,
sleigh rides and limited fly-fishing

Rates: Anglers - $3,200 6 night/5
day (guides included); non-
anglers - $1,600; both rates
include 3 gourmet meals; lodging
and all beverages, including
liquor

THE MARVINE RANCH

The Marvine Ranch is the sister-property to the
famous Elk Creek Lodge. The Wheeler family
opened the Marvine to create the ultimate full-
service private guest ranch. Although the fishing is
as good as that offered through Elk Creek (they
share many of the same waters), the Marvine has a
host of other activities for non-anglers and they
offer a winter program of skiing and
snowmobiling. They have their own lighted
downhill slope, and Aspen, Vail and Steamboat
Springs are only 2 to 2-1/2 hours away. In the
summer, enjoy the equestrian program featuring
gentle horses and thousands of acres of trails. Day
trips to Aspen and Vail are popular with the
guests, as is exploring the hiking trails to see the
plethora of wildflowers or just lounging around
the ranch, watching the wildlife.

The Marvine accepts only complete parties of 8 to
10, creating an intimate atmosphere for family
getaways or small business retreats. It's perfect for
a group of serious anglers who join up with the
Elk Creek angling rotation that includes fishing on
the White River, Trapper's Lake, Marvine and Elk
Creeks and other small spring-fed ponds in the
valley.

Always a highlight at the ranch is a feast prepared
by Executive Chef Bobby Castaldo. A dedicated,
enthusiastic young man, Chef Castaldo has over

six years of professional experience as well as an honors degree from The Western Culinary Institute. His dishes are exciting and bold and are grounded in classical French cooking techniques. He has spent a great deal of time and effort developing partnerships with food purveyors and as a result secures the freshest ingredients, including fresh-caught Pacific fish flown into the ranch the same day it is caught. He has expertise in many cuisines but is especially proud of his knowledge of game cooking. He placed second in a wild game culinary competition against some seasoned Aspen Chefs — no small feat. Before you even get to the ranch, Chef Castaldo will have already discovered your culinary preferences and will plan a week's worth of tantalizing fare worthy of a 4-star restaurant. Chef Castaldo has created this six-course menu for this book, and after you have sampled some of these items I think you will agree that he has phenomenal talent and imagination. You are likely to hear his name again.

B. Hillis

Dinner Menu

SAUTÉED FOIE GRAS WITH A PORT
WINE SAUCE AND WILD
MUSHROOM DUXELLES

BABY ROMAINE AND MAPLE
VINAIGRETTE WITH ROQUEFORT,
TART CHERRIES AND WALNUTS

SAVORY SAFFRON
LOBSTER BISQUE

GRAPEFRUIT CHAMPAGNE SORBET

MARINATED PAN-SEARED ✪
FRENCH LAMB CHOPS
WITH WILD BLUEBERRY
DEMI-GLACE

ROASTED GARLIC AND ✪
ROSEMARY POTATO MASH

BRAISED ASPARAGUS ✪
IN A LEMONGRASS BROTH

CARAMELIZED ONION LOAF

ROBERTO'S CHEESECAKE
WITH GRAND MARNIER AND
RASPBERRY COULIS

✪ Recipe Included

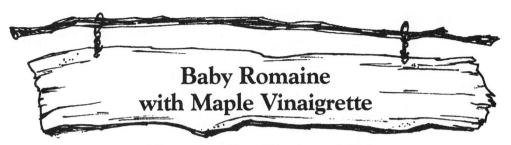

Baby Romaine
with Maple Vinaigrette

...and Roquefort, Tart Cherries and Walnuts

This is an absolutely gorgeous salad, outdone only by the fabulous fresh taste. The Maple Vinaigrette is so delicious I could almost drink it. Chef Castaldo calls for Roquefort blue cheese, but you can use another type if you prefer. If you can't locate baby romaine, buy hearts of romaine instead.

---

4 SERVINGS

Salad:

12 heads Baby Romaine lettuce
1/2 cup crumbled Roquefort cheese, (or your favorite blue cheese)
24 each tart dried cherries
24 each walnut halves, toasted

To prepare salad: Slice lettuce into bite size pieces and divide among 4 chilled salad plates. Sprinkle with crumbled blue cheese, cherries and walnuts. Drizzle 1 to 2 tablespoons of dressing all over salad and plate.

Dressing:

1/2 cup cider vinegar
2 tablespoons maple syrup
1-1/2 teaspoons honey
1/4 cup walnut oil
1/4 cup vegetable oil
Salt and white pepper to taste

Whisk the vinegar, syrup and honey together. Slowly drizzle in walnut oil while whisking. (I prefer doing this in a blender to get a really good emulsion.) Adjust seasoning with salt and pepper.

Marinated Pan-seared French Lamb Chops

with Blueberry Demi-Glace

Is it polite to gnaw the meat off the bone? I should have asked this question before I licked these bones clean! I don't even like lamb that much. This marinade coupled with a fresh cut of meat makes a fabulous gourmet treat. The wild blueberry sauce is unique and delicious, too.

4 SERVINGS

Marinade:

2 cups soy sauce
1 tablespoon Worcestershire sauce
1 teaspoon minced ginger
2 tablespoons minced garlic
1/4 cup Madeira
4 tablespoons Italian dried herbs*
4 tablespoons honey
1 teaspoon onion salt
1/2 cup water
4 (1-pound) lamb racks, 4 or 5 ribs each, bones cleaned (ask your butcher to do this for you)
Salt and pepper to taste
1 cup Wild Blueberry Demi-Glace (recipe follows)

Whisk marinade ingredients together. Add chops and massage until well-coated. Set aside for 30-45 minutes. Remove chops from marinade. Preheat oven to 400°. Heat sauté pan until very hot. Add chops and brown on all sides. Place chops in an ovenproof pan and place in oven for 5 to 10 minutes, or until meat is medium-rare (135°). Let chops rest 5 minutes before carving. To serve, cut between ribs and display chops cut sides up. Drizzle blueberry sauce on lower part of chop and on plate.

Wild Blueberry Demi-Glace:

1 cup demi-glace**
1/2 cup wild dried blueberries
2 teaspoons sugar
1 teaspoon butter

Heat demi-glace and add blueberries and sugar. Simmer until blueberries are reconstituted, about 15 minutes. Remove from heat, add butter and swirl until melted. Salt and pepper to taste if desired.

If you don't have an Italian herb mix handy, substitute 1 tablespoon each of dried basil, oregano, thyme, and marjoram.

**Demi-glace is a concentrated veal or beef stock that has been reduced over a number of hours to a gelatinous stage. You can order demi-glace from More Than Gourmet. See "Sources" section on page 236 for details. One (1.5-ounce) tin of Demi-Glace Gold® makes 1 cup of demi-glace.*

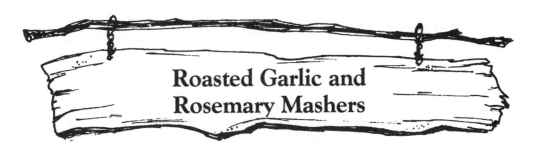

Roasted Garlic and Rosemary Mashers

Of all the garlic mashed potato recipes I've tried, this has to be one of my favorites. The secret is the cream cheese. You can't really taste the flavor of cream cheese, but what it does for the texture is undeniably profound. I think the rosemary is just the right amount, not overpowering, but not so subtle you don't know what herb it is. These spuds are the perfect companion to Chef Castaldo's Marinated Lamb Chops.

6 SERVINGS

1 head of garlic, roasted*
2 sprigs of rosemary, leaves only, stems discarded
5 baking potatoes, about 2-1/4 pounds
1/2 cup butter (1 stick)
1/4 cup heavy cream
1/4 cup sour cream
1/4 cup cream cheese
Salt and pepper to taste

Smear roasted garlic into a paste. Peel and cut potatoes into 1/2 inch cubes and boil until tender, approximately 15 to 20 minutes. Drain the potatoes and put into a mixer with whip attachment and add the rest of the ingredients. Mix until well-combined. For a fancy presentation, run the potatoes through a ricer and put in a pastry bag with a star tip and pipe onto plate. Garnish with an edible flower.

*To roast garlic, see page 11.

Braised Asparagus
in a Lemongrass Broth

This is a haunting combination of flavors and I love the subtle undertones of lemon and ginger. The broth is dainty enough to enhance but not overwhelm the delicate nature of asparagus. Delicious!

4 SERVINGS

Lemongrass Broth:
2 stalks lemongrass
4 cups chicken stock*
1 teaspoon minced ginger

1 pound fresh asparagus
Salt and pepper to taste

To make broth, cut lemongrass stalks into 2" pieces and pound with meat cleaver to release oils. Add stalks and ginger to chicken stock. Simmer for 30 minutes. Trim and peel (if necessary) asparagus and add to broth. Cook 3 or 4 minutes, until crisp-tender.

*Concentrated chicken stock is available through More Than Gourmet. See "Sources" section on page 236 for details. One (1.0-ounce) tin of Fond de Poulet Gold® makes 5 cups of chicken stock.

The Home Ranch

P.O. Box 822
Clark, CO 80428
(970) 879-1780
Website: http://
www.homeranch.com
E-mail: hrclark@homeranch.com

Season: Summer — June through beginning of October

Winter — mid-December through March

Capacity: 40-50

Accommodations: 8 charming cabins of various sizes, 6 rooms in the main lodge; Relais & Chateaux member and Mobil Four Star property

Activities: horseback riding; hiking; fly-fishing; hayrides; barn dances; rodeos; cookouts; swimming; children's program

Rates: Weekly, per couple - $3,230 and up; lodging is charged more by the accommodation than by the person; wide variety in choice of rooms so call the ranch for rates to fit your specific needs

There is a reason they call it The Home Ranch. When you first arrive, you feel like you've been welcomed into someone's home as a special guest. By the time you leave, you feel like it's your home, and you don't want to go. Although The Home Ranch is a working ranch, and you have an abundance of "western" activities in which to participate, it feels more like a luxurious mountain resort. Whether you choose the comfort of a lodge room close to the hub of activity, or a private cozy cabin tucked among the aspens, the beautiful antiques, the stunning original artwork and cozy down comforters beckon you to stay inside. Mother Nature coaxes you outside with magnificent views, crisp, clean mountain air and a flurry of western adventures waiting to be tamed.

Located high in the northern end of Elk River Valley, north of Steamboat Springs, getting to The Home Ranch is a lesson in relaxation. The drive up the valley is surreal, as purple-blue mountains enfold you and the wind-kissed meadows, dotted with cattle, sheep and llamas seem to nod their approval of your arrival. Passing through the quiet town of Clark is another reminder that your hectic city life is another world away. As you come over the last hill in the driveway, a grassy meadow with grazing horses greets you as you gaze upon the handsome lodge in the distance. A tiny voice in the back of your head whispers "You're home."

Horsemanship classes await the novice and experienced rider alike and children have their own program geared toward their special needs. Miles and miles of trails through aspen and pine forests and wide-open wildflower meadows offer breathtaking views and spectacular photographic opportunities. In the winter, these same trails are meticulously groomed for cross-country skiing. Fly-fishing on the Elk River with an experienced guide/instructor is a great way to spend a cool, crisp morning or a warm sunny afternoon as the next hatch of the caddis fly break over the swirling pools tucked along the river's edge. Hiking through the ranch property is an exhilarating experience and often yields fascinating discoveries of western birds, a plethora of colorful wildflowers and perhaps a glimpse of wildlife in their natural habitat.

Steamboat Springs is a short distance away and provides trendy shopping all year and fabulous downhill skiing in the winter. Take advantage of the ranch's daily shuttle to the slopes. The Mount Zirkel Wilderness, a hundred thousand acres of roadless backcountry, is in close proximity for great hikes. Beautiful summer horseback rides or winter wonderland ski trails await you in the Routt National Forest. But don't feel you have to leave the ranch. Many guests stay put and enjoy the abundant pleasures of the ranch life. At the top of the list of pleasure activities is feasting on Chef Clyde Nelson's gourmet cuisine.

The food at The Home Ranch is straightforward yet subtly sophisticated. Blending classical training with a ranch persona gives birth to an eclectic cuisine that is both traditional and cutting edge. Overriding any decision Chef Nelson makes with regard to the menu is its freshness and variety. A communal environment is also important to convey The Home Ranch's tradition of "feeling at home." To take advantage of the awesome weather in the summer, meals are often rotated from inside the lodge dining room to outside in a circle around a huge campfire or a sunset picnic area overlooking Elk River Valley. Regardless of where the meal is served, you can bet that it will be fresh, delicious and worthy of a fine dining establishment. Proof that the food is noteworthy are the dozens of reviews and features produced by Bon Appetit, Food and Wine, Travel and Leisure and cookbook entries in the Spirit of the West and Great Chefs of Colorado cookbooks.

Chef Nelson, who has over 20 years of experience in the industry and 8 years with The Home Ranch, has assembled a team of culinary professionals that would make any restaurant jealous. The talented crew, in tandem with a fresh herb garden, local produce and access to the finest Colorado lamb and game and ranch-raised pork, meld together to create some of the finest food you've ever tasted. Just try some of the recipes Chef Nelson has graciously shared with us and see why The Home Ranch has received such rave reviews from food critics across the country.

Breakfast Menu

FRESHLY SQUEEZED ORANGE JUICE

PLATTER OF SEASONAL FRUITS

BREAKFAST BURRITO IN A HOMEMADE FLOUR ✪
TORTILLA WITH RED AND GREEN CHILE SAUCES

RANCH-RAISED PORK SAUSAGE PATTIES

BLUEBERRY STREUSEL MUFFINS ✪

Dinner Menu

QUESADILLAS WITH BRIE, MANGO & POBLANOS

GRILLED MARINATED SWORDFISH ✪
WITH MANGO SALSA
Or

GRILLED RANCH-RAISED AND BRINED PORK RACK
WITH COLORADO PEACH CHUTNEY

MARINATED GRILLED VEGETABLES
WITH SMOKY TOMATO VINAIGRETTE

WARM APPLE BERRY CRISP
WITH CARAMEL GINGER ICE CREAM ✪

✪ Recipe Included

Breakfast Burrito in a Homemade Flour Tortilla

with Red and Green Chile Sauces

Not only is this dish delicious, it's very attractive with the color-contrasting sauces. It's a flavor-extravaganza as well, with different taste sensations popping out in all directions. When I think of a southwestern breakfast, this is what comes to mind. The best part is the homemade flour tortilla, which is easy to make and oh so much better than store-bought ones. It's a hearty fare, and the ranch serves this with refried black beans. Can't imagine eating again until dinner!

4 SERVINGS

1 tablespoon vegetable oil
8 eggs
2 tablespoons heavy cream
Pinch of salt
1/2 cup grated sharp cheddar or
    Monterey Jack cheese
1 avocado, peeled and chopped

5 sun-dried tomatoes, rehydrated and
    chopped
1 scallion, thinly sliced
1 jalapeno, seeded and finely chopped
1/2 cup each red and green chile sauce
4 tablespoons sour cream
1 tablespoon chopped cilantro
4 (8") flour tortillas

Heat an iron skillet or griddle with a thin film of oil. Beat eggs with the cream and salt. In your favorite egg pan, scramble eggs until soft and still slightly moist. Fold the avocado, sun-dried tomatoes, scallion and jalapeno. Set aside and keep warm over a warm water bath.

Place a tortilla in the hot skillet or griddle and heat 30 seconds then flip over. Spread 2 tablespoons of cheese over the tortilla and allow it to melt. Add 1/4 of the scrambled egg mixture across the center of the tortilla; roll up and let sit in skillet for 30 seconds. Place onto a warm plate and ladle 2 tablespoons of the red sauce on one end and 2 tablespoons of the green sauce on the other end. Top with a dollop of sour cream and sprinkle with cilantro. Keep warm and continue building the rest of the burritos in the same manner.

Flour Tortillas

Because these are so easy and taste best fresh, make these just before you make the burritos. Rolling the tortillas into almost perfectly round shapes is easy once you get the hang of it. Follow the Chef's directions and you'll be cranking these out in no time. The only trouble you'll have is saving them for the burritos instead of scarfing them as they come off the griddle.

YIELD = 12-14 (8") TORTILLAS

2 cups all-purpose flour
2 teaspoons sugar
1-1/2 teaspoons baking powder
1-1/2 teaspoons salt
5 tablespoons vegetable shortening, very cold
3/4 cup hot water

Stir flour, sugar, baking powder and salt in a mixing bowl. Cut in vegetable shortening with a pastry cutter or your fingers. Add hot water and stir together to make a soft dough. Cover with plastic wrap and let stand 15 minutes. Shape dough into golf-ball size balls and dip into flour. Pat dough into a flat round disk with your palm. Using a thin rolling pin, roll each tortilla 2 times in one direction, making an oval shape, then turn the dough 90° and roll 2 more times. Turn 90° and roll 2 times and repeat the process until you get an 8" almost perfectly round tortilla.

Heat a seasoned iron skillet or griddle over medium-high heat. When heated, cook the tortilla on one side until bubbles rise to the surface (about 1 to 2 minutes). Flip tortilla over and cook another 1 to 2 minutes, until lightly browned. Stack cooked tortillas between paper towels to keep them from drying out.

Green and Red Chile Sauce

YIELD = 1 CUP

Green Chili Sauce:
4 poblano chiles or
5 or 6 Anaheim chiles with 1 or 2
    jalapenos
3 cloves garlic
1/4 to 3/4 cup water

1/2 teaspoon ground Mexican oregano
1/4 teaspoon ground cumin
2 tablespoons fresh cilantro leaves
1/2 teaspoon salt
1/4 medium onion, roughly chopped
1 tablespoon olive oil

Roast the chiles and garlic in a 350° oven for 35 to 40 minutes. Put the chiles in a bowl, cover with plastic wrap and let them "steam" for 10 to 15 minutes. Finely chop garlic and set aside. Peel the chiles and place in a food processor or blender along with the remaining ingredients and process to a medium consistency, but do not puree. Warm in a saucepan before serving.

Red Chile Sauce:
2 roma tomatoes
2 ounces dried red chiles (such as New
    Mexican, ancho and cascabels)
2 cups water
1-1/2 teaspoons olive oil

1/2 small onion, chopped
1 clove garlic
1/2 teaspoon ground cumin
1/2 teaspoon ground Mexican oregano
1/4 teaspoon salt
1 tablespoon vegetable oil

Core tomatoes and in a cast iron skillet over high heat, blacken the tomatoes on all sides. Remove stems and seeds from the dried chiles and place them on a sheet pan. Dry roast them in a 350° oven for 2 to 3 minutes, or until they begin to emit an aroma (careful not to cook to long or they will burn). Rehydrate the chiles in boiled water for 20 minutes (reserve chile water). Heat the olive oil in a skillet and add the onion. Cook over medium heat until golden brown, 5 to 10 minutes. Place blackened tomatoes, rehydrated chiles and all other ingredients in a food processor or blender. Add some of the chile water and process until blended and of sauce consistency, adding more chile water as necessary. Heat the vegetable oil in a skillet and add the sauce, cooking 3 to 5 minutes and stirring constantly, adding water if necessary. (I strained the sauce even though the recipe doesn't call for it. I don't like the bits and pieces of the tomato and chile skins left in the sauce after processing.)

Blueberry Streusel Muffins

In a side by side taste test of these and 3 other anonymous muffins, these won hands down among all the tasters. I think it's the crunchy topping that sets this apart from most other muffins (not to mention it's also covered with a glaze). Besides the crunch, the topping is sweet with a noticeable tang from the lemon peel and the lemon glaze. These muffins are extremely moist and very berry-tasting. You can use frozen blueberries instead of fresh, but your muffins will be purple.

YIELD = 15-18 MUFFINS

1 cup sugar
1 cup vegetable oil
4 eggs, room temperature
1 teaspoon vanilla
3 cups all-purpose flour
1 tablespoon baking powder
1 teaspoon baking soda
1/4 teaspoon salt
1 cup buttermilk, room temperature
2 cups fresh blueberries

Topping:
1/2 cup chopped pecans
1/2 cup dark brown sugar
1/4 cup all-purpose flour
1 teaspoon cinnamon
1 teaspoon grated lemon peel
2 tablespoons melted butter

Glaze:
1/2 cup powdered sugar
1 tablespoon lemon juice

Preheat oven to 375°. Mix all topping ingredients together until it resembles coarse crumbs. Set aside.

Beat sugar and oil and eggs for 25 seconds with a wooden spoon. Add vanilla and stir. Add flour, baking powder, baking soda and salt and stir. Add buttermilk and stir just to combine. Fold in blueberries. Fill greased muffin tins 2/3 full. Sprinkle with topping. Bake 15-18 minutes. Remove from oven and cool 5 minutes. Remove from pan to a cooling rack with a sheet pan underneath. Mix powdered sugar with lemon juice until smooth. Drizzle over muffins.

Grilled Swordfish with Mango Salsa

This is a dynamic marinade, perfect for a firm fish. It's citrusy and subtlety oriental. If you like swordfish, you'll love this. If you don't, I think you will like this.

6-8 SERVINGS

6 to 8 (6-ounce) fresh swordfish steaks
Marinade:
1/2 cup fresh squeezed orange juice
    (about 1 orange)
1/4 cup fresh squeezed lime juice
    (about 3-4 limes)
1/4 cup rice wine vinegar

3 tablespoons peanut oil
1 tablespoon sesame oil
2 tablespoons fresh ginger
3 cloves garlic, sliced into ovals
2 green onions, sliced thinly
2 tablespoons chopped fresh cilantro

Combine all ingredients and whisk vigorously. Pour marinade over fish about 1 hour before grilling. Serve with Mango Salsa (recipe follows).

Mango Salsa

1 mango, peeled and chopped
1/2 pineapple, cored, peeled and
    chopped
1/2 small jicama, peeled and chopped
1/2 red bell pepper, chopped
1/2 yellow bell pepper, chopped
1 tablespoon fresh grated ginger
2 serrano chiles, seeded and chopped (or
    1 jalapeno, seeded and chopped)

2 cloves garlic, finely chopped
8 leaves fresh basil, chopped
2 tablespoons chopped fresh cilantro
1 small bunch fresh mint, chopped
1 teaspoon soy sauce
2 tablespoons peanut oil
1 tablespoon sesame oil
Juice of 2 limes
Salt and pepper to taste

Mix all ingredients together. Can make 1 day ahead to allow flavors time to blend. Heat grill to medium heat (350°). Remove swordfish from marinade and place on hot grill. Grill for 5 to 7 minutes per side, or until desired doneness. Remove and place on a warm plate. Top with 1/4 cup Mango Salsa.

Ginger Caramel Ice Cream

While you probably wouldn't sit down to a whole big bowl of this like you would, say, chocolate chip ice cream, this ginger-caramel combination is a perfect accompaniment to a warm apple pie. It's rich, smooth and creamy. Can't ask much more from an ice cream.

YIELD = 1 QUART

1 cup sugar
1 tablespoon peeled, thinly sliced fresh ginger
1 tablespoon water
2 cups heavy cream, warm
2 cups milk, warm
7 egg yolks
1 tablespoon sugar

Heat the cup of sugar, ginger and 1 tablespoon of water over medium heat in a heavy-bottomed saucepan. Cook until the sugar turns dark brown. (It may crystallize, which is fine, just continue cooking until it re-melts and turns brown. It took about 18 to 20 minutes for me.) Very carefully pour in the warm cream and milk and stir with a whisk. (If the caramel seizes up into hard chunks, continue cooking over moderate heat, stirring until it all melts.) Bring to a bare boil and remove from heat. Beat the egg yolks with 1 tablespoon sugar. Add 2 or 3 tablespoons of the hot caramel mixture into the egg mixture and stir. Add a few more tablespoons of caramel mixture into eggs and stir. Pour tempered egg mixture into caramel mixture and cook over low heat until mixture slightly thickens (coats the back of a spoon without dripping), about 2 or 3 minutes. Strain into a bowl and quickly chill in an ice bath. When mixture is cool to the touch, cover and transfer to the refrigerator. Chill for 1 or 2 hours. Transfer mixture to ice cream machine and freeze according to manufacturer's directions. Store in an airtight container in the freezer for up to 3 weeks.

Vista Verde Ranch

P.O. Box 465
Steamboat Springs, CO 80477
(970) 879-3858
(800) 526-7433
Website: http://
   www.vistaverde.com
E-mail:
   103573.3551@compuserve.com

Season: June through September;
December through March

Capacity: 30+ in summer;
20+ in winter

Accommodations: Quiet
mountain log cabins that are
rustic on the outside and elegant
on the inside and handsomely-
appointed new lodge rooms with
private balconies

Activities: Summer — horseback
riding; fly-fishing; hiking; rock
climbing; rafting; rodeos;
kayaking; mountain biking; hot
air ballooning; children's
program; Winter — skiing,
snowshoeing; sleigh rides,
horseback riding; sledding; dog
sledding; fly-fishing; ice-climbing

Rates: $1,550-$1,850 weekly per
person; includes lodging, all
meals and most activities;
children's rate $500 less

The new American West, it's where raw territorial
beauty abounds in every direction, where everyone
is a neighbor, even if they live 500 acres away and
where roughing it is not so rough. It's retaining
the heritage and spirit that made it once wild. It's
Vista Verde, the epitome of the new American
West. Tucked away high up in the Elk River
Valley, just 25 miles north of Steamboat Springs,
Vista Verde is a working hay and cattle ranch and
a Mobil Four Star guest ranch. Many things make
Vista Verde unique including the fact the cattle
operation peacefully co-exists with the guest ranch
operation. In fact, adventurous guests help in the
semi-annual cattle round-ups and other ranch
chores. The 1 to 1 staff-to-guest ratio is
phenomenal. The vast array of activities both in
summer and winter simply cannot all be
accomplished in a one-week visit. There is one
television and one phone at the ranch, but you
won't find either unless you ask. The ranch wants
you to forget about the outside world for a while
and engage in total rest and relaxation, be as active
as you want to be, let yourself be pampered.

Vista Verde has all the prerequisites of a great
western guest ranch, including an abundance of
breathtaking scenery, a secluded, serene habitat, a
strong horse program and fantastic fishing and
hiking opportunities. And, and, and... The
activities list is one of the longest I've seen. They
even offer ice-climbing in the winter. However, it's
the attentive staff, the exceptional service, the first-
class accommodations and the superior cuisine
that elevates Vista Verde from a great ranch to an
extraordinary vacation destination.

The culinary experience at Vista Verde is as diverse and appealing to all types as are the activities. The innovative, creative talents of Chef de Cuisine Jonathon Gillespie and his team will pleasantly surprise well-traveled guests. A Culinary Institute of America graduate, Chef Gillespie honed his culinary skills at several Hilton fine-dining properties in North Carolina before coming West. As a result, the Vista Verde dining experience is a fusion between the fundamental elements of classic cuisine and the casual ambiance of western fare. It's a fabulous combination and one you'll not likely forget. The staff says the food "walks a fine line between ranchy and fancy." Where else would you find a burrito and a smoked salmon white cheddar cheese sauce on the same menu?

Working in a fabulous new kitchen, designed for chefs by a chef (this is unusual, believe me), the culinary crew is empowered to be creative and to experiment with new dishes. This leaves them free to develop a western flair for old classics as well as invent new culinary delights. The herb and vegetable gardens just outside the kitchen provide daily inspiration to the chefs. Chef Gillespie and his capable crew, including Sous Chef Charles McGlynn, have developed a stable of "lighter" cuisine to give guests an option of tempered self-control versus the typical vacation mode of abandoned gluttony!

Chef Gillespie has shared two menus with us that are representative of both the diversity and balance of color, texture and flavor offered. You can count on two things when you settle in for a meal at Vista Verde. One, you have a choice from an array of tempting dishes, and two, your taste buds will dance and applaud your selection, whatever that might be.

Breakfast Menu

CINNAMON RAISIN BREAD ✪
WITH MAPLE GLAZE

CORNMEAL WAFFLES ✪
WITH SMOKED SALMON AND A
WHITE CHEDDAR CHEESE SAUCE

Dinner Menu

FRESH BAKED BLACK PEPPER
POTATO BREAD

BEEF TENDERLOIN ✪
WITH AN APPLE SMOKED BACON,
WILD MUSHROOM AND GRILLED
RAMP COMPOTE

WILD RICE
WITH TOASTED PECANS
AND BLACK GRAPES

HARICOT VERTS

MOLTEN MEXICAN ✪
CHOCOLATE ECSTASY

✪ Recipe Included

Cinnamon Raisin Bread
with Maple Glaze

I like how tall this bread is. It's the kind of bread Julia would photograph for her baking book. Tender and moist, this bread is delicious right out of the oven or even couple of days later. The oatmeal helps make it moist, but you really can't tell there is oatmeal in the finished loaf. The maple glaze adds a complementary flavor to the cinnamon and just the right sweetness. I used the dough hook on my stand mixer, but you can make this dough by hand if you prefer.

YIELD = 1 VERY TALL LOAF

1 cup warm water (110°)
1/4 cup maple syrup
1/2 tablespoon yeast
5 cups all-purpose flour, more or less
1-1/2 cups oatmeal, hot
1 tablespoon + 1 teaspoon butter

1 teaspoon salt
2/3 cup raisins
2 teaspoons cinnamon

1 cup powdered sugar
3 tablespoons maple syrup

In a mixing bowl, stir together warm water, maple syrup and yeast. Set aside for 5 to 10 minutes. Meanwhile, heat oatmeal and add butter. Stir to melt butter. When yeast mixture is bubbly, add 3 cups of the flour, oatmeal, salt and raisins to the yeast mixture. Mix on low speed until incorporated. Keep adding the flour in 1/2 cup increments until the dough comes together and is only slightly tacky to the touch. The dough will be very soft and somewhat sticky. Place dough in a greased bowl, cover with plastic wrap and place in a warm place to rise until double in size (about 40 to 60 minutes).

Spray a 9" X 5" loaf pan with non-stick cooking spray. Turn the dough out onto a lightly floured surface and knead the cinnamon in by sprinkling a little on the top, kneading and turning the dough and repeating the process until all the cinnamon is used. Place the dough in the prepared loaf pan and place in a warm place to rise. When the dough has risen about 2" above the pan, put in a preheated 350° oven and bake for about 30 minutes. The crust will be golden brown. Remove from pan and place on a cooling rack.

Once the bread has cooled, mix the powdered sugar with the 3 tablespoons of maple syrup. Add 1 or 2 tablespoons of warm water to thin to a pouring consistency. Pour glaze all over the top of the bread, letting it run down the sides.

Cornmeal Waffles
with Smoked Salmon

and White Cheddar Cheese Sauce

How utterly unique! Perfect for a sophisticated brunch or very special guests, these waffles deserve center stage. The delicate texture of the cornmeal is superb with the exquisite smooth, velvety sauce. The small cuts of the peppers make the sauce attractive and the ancho chili adds a warm depth. The elaborate prep work for the sauce is more than worth the effort. To save time, do all the preparation for the sauce the day before by chopping all the vegetables and grating the cheese. You can even partially cook the sauce the day before as stated in the directions.

8 SERVINGS

Sauce:
2 tablespoons vegetable oil
1-1/2 tablespoons finely chopped shallot
1-1/2 tablespoons finely chopped garlic
1/2 red bell pepper, finely chopped
1/2 yellow bell pepper, finely chopped
1/2 green bell pepper, finely chopped
1 ancho chile, rehydrated and finely
 chopped
1 cup chicken stock*

1-1/2 cups heavy cream
1-1/2 cups grated white cheddar cheese
8 ounces sliced smoked salmon
1 tablespoon finely chopped cilantro
1/4 bunch basil leaves, finely chopped
Salt and pepper to taste

8 cornmeal waffles (recipe follows)
2 roma tomatoes, peeled, seeded and
 finely chopped

In a large sauté pan or medium saucepan, heat oil over medium heat. Add shallot and garlic and sauté 1 minute. Add bell peppers and ancho chile. Cook, stirring occasionally for 5 minutes. Add stock and bring to a boil. Reduce heat and simmer until liquid is reduced by half. (At this point, you can remove from heat, cool and store in the refrigerator for 1 to 2 days.) If refrigerated, warm sauce before proceeding with recipe. Add cream and bring to a boil for 5 to 7 minutes. Add cheese and stir to melt cheese. Take half of the salmon slices and cut into thin strips, 1" to 2" long. Add to sauce. Add cilantro and basil. Season with salt and pepper to taste. Spoon sauce over cornmeal waffles and garnish with remaining salmon slices and chopped roma tomatoes. (You can cut the remaining salmon into strips and roll into decorative curls.)

*Concentrated chicken stock is available through More Than Gourmet. See "Sources" on page 236 for details. One (1.0-ounce) tin of Fond de Poulet Gold® makes 5 cups of chicken stock.

Cornmeal Waffles

YIELD = 8 WAFFLES

1-1/2 cups cornmeal
4 tablespoons all-purpose flour
2 teaspoons sugar
1 teaspoon baking powder
1 teaspoon salt

1/2 teaspoon baking soda
4 eggs, lightly beaten
2 cups buttermilk
1/2 cup butter (1 stick), melted

Combine first 6 ingredients, cornmeal through baking soda, in a large bowl and stir. In a separate bowl, mix eggs and buttermilk together. Pour egg mixture over flour mixture, add melted butter and stir just to combine. Heat a waffle iron and cook waffles according to manufacturer's directions.

B. Hillis

Black Pepper Potato Bread

Wow! This bread's got a nice kick. I'm sure I'm supporting an entire pepper farm because I buy so much black pepper so, obviously, I'm going to like this bread. If you don't like black pepper, either skip this recipe or for the heck of it, give it a try. Bread is fun to make, and you can give it as a gift if you don't want it. Mail it to me. I love it.

YIELD = 2 LOAVES

2 cups warm water
1 tablespoon honey
1 tablespoon yeast
5-7 cups all-purpose flour
1-1/2 cups hot mashed potatoes (about 2 small or 1 large potato)
6 tablespoons butter
2 teaspoons salt
1-1/2 tablespoons black pepper

In a mixing bowl, add the water, yeast and honey. Stir and let stand 5 to 10 minutes. Using a dough hook, begin mixing the dough by adding 3 cups of flour to the yeast mixture. Add the potatoes, butter and salt. Add black pepper and continue to add flour until the dough is only slightly tacky to the touch. Place dough in a large greased bowl, cover with plastic and place in a warm place to rise. When double in size (about 40 to 60 minutes), divide dough in half and shape into 2 loaves. Place dough in 2 greased loaf pans. Let the dough rise again until about 1" to 2" above loaf pan (about 20 to 40 minutes). Bake in a 350° oven for 25 to 30 minutes. Remove from pans and cool on a rack.

Grilled Beef Tenderloin

with an Apple Smoked Bacon,
Wild Mushroom and Grilled Ramp Compote

Oh my! This dish is too fabulous. The beef tenderloin is drenched in a
haunting molasses marinade that is simply ambrosial. When grilled, the
marinade forms a sublime crust on the tenderloin that at first blush seems
sweet, then unfolds as tart and then back to sweet again. If this sounds
confusing, you'll just have to try it to see what I'm talking about. I don't
think I've ever had a better piece of tenderloin. Everything flows together
like a symphony, from the tender beef to the flavor-complex compote, to the
wild rice dish Chef Gillespie pairs with this spectacular dish. If it sounds like
I'm gushing it's because I am. I can't tell you enough how good this is. You'll
have to see for yourself.

4 SERVINGS

Molasses Marinade:
1 cup molasses
2 tablespoons balsamic vinegar
2 tablespoons black pepper
1 tablespoon finely chopped garlic

1/2 tablespoon finely chopped shallot
2 teaspoons fresh grated gingerroot
1 teaspoon dried thyme
1 teaspoon dried marjoram
1/2 teaspoon red pepper flakes

Mix all ingredients together and let rest 1 hour for flavors to blend.

Beef Tenderloin:
1-1/4 pounds whole beef tenderloin
1/2 cup molasses marinade

Marinate tenderloin in molasses for 30 minutes up to 4 hours. Heat grill to medium-high
heat (375°-400°). Grill tenderloin about 5 minutes on each side, for a total of 20 to 25
minutes for medium rare (130°). Remove from grill and let rest 5 minutes before slicing
into 3/4" medallions. Serve with Wild Mushroom Compote (recipe follows).

Apple Smoked Bacon, Wild Mushrooms and Grilled Ramp Compote

Ramps are wild leeks and much smaller than those you get in the grocery store. If you can't locate them, substitute scallions. Throw these on the grill as you put the beef tenderloin on, and grill until they are slightly wilted, about 5 minutes. You can also use any smoked bacon. Chef Gillespie prefers apple wood for smoking meats.

YIELD = 2 CUPS

1/2 cup molasses marinade
1 cup demi-glace*
1 cup chopped uncooked apple-smoked bacon, about 8 to 10 slices
3 tablespoons bacon fat (will come from cooking the bacon)
2 cups sliced wild mushrooms (such as shiitake, oyster, cremini, morels, etc.)
1 bunch ramps (or scallions) grilled
2 teaspoons cider vinegar
Pinch salt and pepper

Combine molasses and demi-glace in a medium saucepan over high heat and cook until the liquid is reduced by half. (You'll have to take the pan off the heat to judge, as it bubbles and increases in volume over the heat.) Meanwhile, in a large sauté pan, cook bacon until crisp. Remove bacon with a slotted spoon and set aside on paper towels to drain. Remove all but 3 tablespoons of bacon fat. Cook mushrooms in fat for 3 to 5 minutes, or until tender. Add sliced grilled ramps and vinegar. Add bacon and the molasses mixture and cook for 10 minutes on low, or until sauce reaches desired consistency. Place compote on lower 1/3 of beef medallion, letting it run onto the plate.

*Demi-glace is available through More Than Gourmet. See "Sources" section on page 236 for details. One (1.5 ounce) tin of Demi-Glace Gold® makes 1 cup of demi-glace.

Molten Mexican Chocolate Ecstasy

The name is pretty descriptive. It's awesome. I'd rather eat this than dinner if I had to make a choice. The presentation in a cup is unique, too. The top is crunchy, the middle is gooey and the bottom is cake-like. If you practice, you can get all three textures in one spoonful for a great sensation. Of course, it helps if you have a big mouth, like me.

5 SERVINGS

1 to 2 tablespoons butter
1/4 cup granulated sugar

1-3/4 sticks butter, softened
6-1/2 ounces Mexican chocolate*
4 eggs
4 egg yolks
1/2 cup + 1 teaspoon all-purpose flour
1-1/4 cups powdered sugar

Preheat oven to 425°. Rub 4 or 5 oven-proof coffee cups with butter and coat with granulated sugar, pouring out excess sugar. Set aside. Chop chocolate and melt over simmering water in a double boiler. Whisk 1-3/4 sticks butter into chocolate, and then beat in eggs. Mix in the flour and powdered sugar. Pour enough batter into prepared cups to come 1" from top of cup. Bake in 425° oven for 12 to 15 minutes. (The middle should be liquid.) Serve warm.

If you cannot locate Mexican chocolate, substitute semi-sweet chocolate and add 1/2 teaspoon cinnamon and 1/4 teaspoon almond extract to chocolate after it's melted.

Wit's End Guest Ranch & Resort

254 County Road 500
Bayfield, CO 81122
(970) 884-4113

Season: Year-round; American plan April through December; optional European plan January through March

Capacity: 216

Accommodations: 34 luxury class historic log cabins, all with fireplaces, rock hearths and covered decks; additional space for corporate groups in Silver Streams and Trout House Lodges located on property

Activities: Trail riding, arena riding; guided hiking; bicycling; tennis; swimming; fishing in private ponds, lake or rivers; children's program; nightly entertainment and more; full winter program of skiing, snowshoeing, sledding, snowmobiling

Rates: $1,840-$2,192 weekly per person

"Luxury at the edge of the wilderness" is how Jim and Lynn Custer, owners of Wit's End Guest Ranch, describe their southwestern Colorado dude ranch. Sequestered in the narrow Vallecito River Valley, Wit's End is surrounded by rugged mountain peaks towering from 12,000 to 14,000 feet. The sprawling 300+-acre ranch is adjacent to the 465,000-acre Weminuche Wilderness in the heart of the two-million-acre San Juan National Forest. Wilderness? Yes. Roughing it? No. The accommodations may be rustic, historic log cabins on the outside, but they are uniquely appointed with the finest furnishings and handsome décor on the inside.

The central gathering place for the ranch is an old barn, constructed in the late 1800's by homesteader John Patrick. Prior to Mr. Patrick's ownership, the ranch was the hideout for the infamous outlaw Dalton Gang. Don't miss the ride up to the original cabin, at 12,000 feet elevation, that the Daltons built as a lookout point. Even earlier, the area was home to Spanish gold miners and French fur trappers. As much as the ranch is rich in history, it is just as replete with modern amenities. The Custers renovated the three-story barn, restoring the original stone floors, huge hand-hewn beams and the enormous rock fireplace. Called the Old Lodge, the "barn" is now home to an antique library, billiards room, dining room and authentic western tavern.

Riding is the mainstay of ranch activities, though there are plenty of non-riding endeavors as well. The ranch is a launching pad for trips to the Mesa Verde Indian Ruins, the historic town of Durango and the famous Silverton-Durango railroad as well as the up and coming Purgatory ski area. Sticking close to the ranch, you can plan multiple activities on the seven-mile-long Vallecito Lake, including fishing, water-skiing and sail-boating. Other than cookouts, hayrides, and trail rides, activities are not structured or scheduled so you can set your own pace, participating in as many or as few ventures as you'd like. The children's program is quite extensive and provides a great opportunity for parents to enjoy time with and without children. Trained counselors, a separate dining area and recreation center offer peace of mind to parents and guaranteed fun for the kids.

Wit's End is the perfect place for a business retreat or a family reunion or even a wedding. The ranch has hosted over 400 weddings since 1988, offering a unique setting and romantic getaway for new couples and an adventurous environment for their extended families. Corporate meeting planners are always pleased with the ranch's facilities and professional staff, making their jobs coordinating business retreats a breeze. The repeat rate is phenomenal among business groups as well as individual families.

An exceptional lodge requires exceptional food, and this is yet another area in which Wit's End excels. Through the course of a week, you'll experience five-course gourmet meals and authentic western cookouts encircling a blazing campfire. You'll dine inside and outside and marvel at the talents of the Wit's End culinary team. A copious kitchen staff headed by an Executive Chef with more than 20 years of experience fashions an unrelenting stream of tempting fare. The Pastry Chef captivates your imagination and palate with mouth-watering desserts and robust breads. The range of cuisine offered is as wide and varied as the terrain.

Escargot en Pasta may share a menu with Lone Star Chicken Fried Steak. Southwestern Chicken Tempura or Veal Piccata or Catfish Dalton or Dover Sole a la Meuniere, and the list goes on. Comfort foods to classic French dishes coexist in harmony and each is presented in a style worthy of the finest restaurant. I couldn't begin to fit an evening's dinner selection on one page. The choices are astounding, all eloquently prepared and presented. I've created a menu from the multitude of choices you are likely to see. I'm inspired by the spectrum of flavors and the artful presentation techniques within these recipes as well as other fare offered by the ranch. I think you, too, will be impressed and pleased with the results.

Dinner Menu

SALMON RAVIOLI ✪
WITH CREAMY BASIL SAUCE

WILD, WILD MUSHROOM SOUP ✪

GRILLED HALIBUT ✪
WITH WIT'S END SALSA

Or

SOUTHWESTERN STUFFED
CHICKEN BREAST WITH GREEN
CHILE SAUCE

APPLE DUMPLING DARLING ✪

✪ Recipe Included

Salmon Ravioli
with Creamy Basil Sauce

Very elegant and classy. The taste is divine, too, especially with the rich creamy sauce. The raviolis are folded in a unique manner that takes a little time. We also used a round biscuit cutter with scalloped edges. While we did save time, they weren't as pretty or unusual. The ranch's presentation is worth a little extra time.

6 SERVINGS

1 (8-ounce) salmon filet	2-3 finely crushed saltine crackers
1 cup heavy cream	1/8 cup shredded Parmesan cheese
1 egg yolk	1 package wonton wrappers
Salt and pepper	Basil leaves for garnish

Poach salmon in lightly simmering water (160°-185°). It will take about 10 minutes for every 1" thickness of salmon. Remove from water and cool. Remove skin and any remaining bones and flake fish meat with a fork. Heat cream in a small saucepan and simmer until it's reduced by half, about 30 minutes. Add flaked salmon and Parmesan cheese. Cook 1 minute, stirring. Beat egg yolk in a small bowl and add a little hot salmon mixture to the egg and stir. Now add the tempered egg yolk to the salmon mixture, stirring constantly. When fully incorporated, remove from heat and stir in crushed crackers and Parmesan. Add enough to shape mixture into a loose ball. Taste and season with salt and pepper. Put mixture into the freezer for 15 to 20 minutes to firm up.

Place one wonton wrapper on a work surface and brush with water. Place 1 tablespoon of salmon mixture in the center and cover with another wrapper. Press tightly around filling. Brush this top wrapper with water. Fold in all 4 edges, toward the center, twice, trimming as necessary to make edges lie flat (see diagram at right). Press a fork on the outer seams to seal and place on a rack to dry, 10 to 15 minutes. Refrigerate or freeze if you are not using right away. When needed, poach 3 raviolis per person in 1 quart of salted water, for 4 or 5 minutes if thawed and 12 to 15 minutes if frozen. These are very delicate, so handle with care to prevent breaking. Place 2 tablespoons of basil cream sauce in a shallow serving bowl, top with 3 ravioli and drizzle with more sauce. Garnish with a fresh basil leaf.

Creamy Basil Sauce:

3 cups heavy cream	1/4 cup or more Parmesan cheese
2 tablespoons chopped fresh basil	Salt and pepper to taste

Heat cream in a medium saucepan and simmer until it's reduced by 1/3 (you'll have 2 cups left), about 30 minutes. Add 2 tablespoons chopped basil and thicken with Parmesan cheese. Season to taste with salt and pepper.

Apple Dumpling Darling

This belongs in a picture book! It is absolutely stunning to look at and heavenly to eat. It's not a difficult dish to make yet it looks like a work of art. My neighbor Alex took pictures of it before he and Rosalie devoured every last crumb. I used the T Cross chocolate sauce and the Elk Mountain's Spicy Apple Cider Sauce to garnish this plate. Whatever sauces you use should be contrasting in flavor as well as color for the full dramatic effect.

6 SERVINGS

6 Granny Smith apples, peeled and cored
4 cups water
5-1/3 cups sugar
1 cup cinnamon sugar mixture (see ingredients listed)
1 cup brown sugar
12 sheets phyllo dough
1-1/2 cups butter (3 sticks) melted

Cinnamon sugar mixture:
2/3 cup sugar
1/3 cup cinnamon
2 teaspoons ground ginger
1/2 teaspoon nutmeg

In a narrow but deep stockpot, bring 4 cups of water and 5-1/3 cups sugar to a boil. Reduce heat to a bare simmer (170°) and stir until sugar is melted. Add whole, peeled and cored apples and poach until almost done, turning often to cook all sides, about 12 to 15 minutes. You will be able to easily stick a fork in the apples about 1/4" to 1/2" before the fork meets resistance. Remove from syrup and cool.

Keep phyllo sheets covered with plastic wrap and a wet towel on top of the plastic, bringing out 1 at a time. Place one phyllo sheet on a work surface and brush liberally with melted butter. Fold sheet in half, bringing short sides together. Place another phyllo sheet on top, with half of the second sheet on top of the first folded sheet. Brush entire second sheet with butter and fold in half. You now have 4 even layers. Butter the top layer. Roll the apple lightly in the cinnamon sugar mixture. Place the apple upright in the center of your buttered 4 layers of phyllo dough. Fill an apple hole with 2 tablespoons of brown sugar. Bring the four corners up over the top of the apple and pinch and twist together, like a drawstring purse. Tuck and smooth the phyllo dough to make it look nice. Place on a baking sheet and repeat process, starting with buttering the phyllo dough, for each apple.

Continued next page

THE GREAT RANCH COOKBOOK

Apple Dumpling Darling (continued)

Preheat oven to 400°. Bake apples for 15 minutes, or until golden brown. Remove from oven. To present the apples, coat a dessert plate with 1/4 cup light-colored sauce (either a vanilla sauce or the Spicy Apple Cider Sauce on page 90). Around the edge of the plate drop dots of a dark or contrasting color sauce, about 1/2" in diameter and about 1/4" apart all the way around. It's important that the dots are on top of the light sauce, so make sure the light sauce is spread to the edges of the plate before you drop the dots. It's much easier if you have your dark sauce in a squeeze bottle to do this. (We used the Chocolate Sauce from T Cross on page 227.) Take a toothpick and put the tip in the center of a dark dot. With the toothpick touching the plate, drag it all the way around the plate and through the center of each dark dot. You will have beautiful hearts, symmetrically circling your plate. Practice a few times so you get the hang of it. When you're happy with your decorating, place a warm apple dumpling in the center of your plate and serve.

Wild, Wild Mushroom Soup

The ranch just calls this Wild Mushroom Soup. I added the other "Wild" to communicate the extreme mushroom flavor. I love the earthy robustness of this delicious soup. You can make it with or without the cream. I prefer the cream version (of course) as it mellows the lustiness of the soup just a bit and adds a luscious richness. I also prefer fresh as opposed to dried mushrooms. Even though the dried mushrooms are reconstituted, the texture is still a little chewy, though the flavor is delicious.

YIELD = 1 QUART

4 ounces fresh wild mushrooms
    (or 2 ounces dried)*
2 shallots, finely chopped
1/2 medium onion, chopped
1/2 cup chopped red bell pepper
3 cups fresh chopped button mushrooms
2 tablespoons butter

1/4 cup all-purpose flour
1 quart chicken stock, heated**
1/2 cup Pernod or other licorice flavored
    liqueur
1/2 teaspoon fresh thyme
2 cups half and half, heated

Melt butter in a large sauté pan over medium heat. Add shallots, onion, red bell pepper and button mushrooms (and fresh wild mushrooms if using). Cook for 5 or 6 minutes and add liqueur. Flame liqueur. After flame dies out, sprinkle flour over mixture and stir constantly until well incorporated and mixture turns a light golden color, about 5 to 7 minutes. Meanwhile, in a large saucepan, heat stock to boiling. Add mushroom mixture, whisking vigorously to incorporate well. (Add the reconstituted wild mushrooms, if using, at this point). Season with fresh thyme. Simmer for 20 minutes, skimming surface if necessary. Season with salt and pepper. If using, add heated half-and-half and adjust salt and pepper if necessary.

If you're using dried wild mushrooms, soak in boiled water for 2 hours and then drain and chop.

**Concentrated chicken stock is available through More Than Gourmet. See "Sources" section on page 236 for details. One Fond de Poulet Gold® (1 ounce) tin makes 5 cups of chicken stock. You need 4 cups for this recipe.*

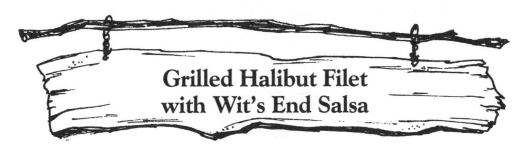

Grilled Halibut Filet with Wit's End Salsa

Halibut is one of my favorite fishes and this simple presentation allows the natural flavor and texture of the fish to shine through. The salsa has a good kick to it and is an attractive color addition to the white fish.

6 SERVINGS

6 (6-7 ounce) halibut filets
Salt
Lemon pepper
1/2 stick butter, melted

1-1/2 cups Wit's End Salsa (recipe follows)
2 tablespoons chopped cilantro
6 lemon wedges

Preheat grill to medium-high heat (375°-400°). Season filets with salt and lemon pepper. Grill halibut to desired doneness, about 5 to 8 minutes per side for medium. (To make cross marks, first lay filet pointing to the 11:00 position on an imaginary clock. Cook for 3 minutes and pick up fish and turn so that the fish, same side, is pointing toward 1:00. Cook for another 3 minutes and flip over to cook other side.)

Brush grilled filet with melted butter and place on a warm plate. Place 1/4 cup salsa on one end of fish and sprinkle entire fish with a 1/2 teaspoon cilantro. Add a lemon wedge.

Wit's End Salsa

YIELD = 1-1/2 CUPS

1 large tomato, chopped
1 green onion, thinly sliced
1 jalapeno, seeded and finely chopped
1/4 cup finely chopped onion
1 cup tomato juice
Salt and pepper

Mix all ingredients in a bowl and cover. Let rest for 2 to 3 hours. When ready to serve, remove with a slotted spoon to drain liquid.

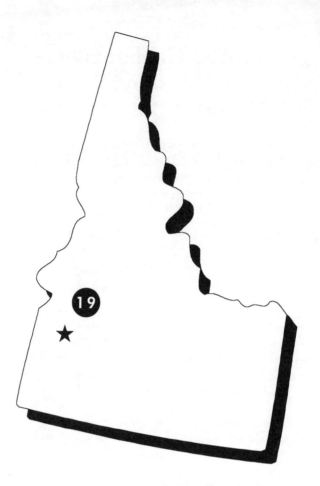

IDAHO

THE GREAT RANCH COOKBOOK

Wapiti Meadow Ranch

H.C. 72
Cascade, ID 83611
(208) 633-3217
Website: http://
guestranches.com/wapiti.htm
E-mail: wapitimr@aol.com

Season: May through October;
January through March

Capacity: 12-14

Accommodations: 3 two-bedroom cabins, 1 one-bedroom cabin, and 2 lodge rooms

Activities: Orvis-endorsed fly-fishing; horseback trail riding; mountain bike riding; hiking; gold-panning

Rates: 6-day weeks $1,000 to $1,980, per adult per week, depending on season

"This is a real treasure. Horseback riding, great food, Orvis-quality fishing, articulate guides and a gracious hostess." This guest comment describes the experience you'll have at the Wapiti Meadow Ranch in Idaho. Owner Diana Swift leaves no detail to chance and prides her ranch on total guest satisfaction. The ranch is located high in central Idaho, 60 miles away from the nearest habitation, in a jeweled meadow that can only be described as breathtaking. The red-roofed lodge and cabins sit amid manicured lawns and flower gardens. Peaceful. Restful. Nearby streams and mountain peaks capture the essence of a relaxing vacation. Don't miss the fly-out trip to the Middle Fork of the Salmon River, deep in the heart of the River of No Return Wilderness for superb fly-fishing and an overnight camp that features filet mignon on the menu.

Diana describes her cuisine as "Hearty Gourmet." I call it fabulous! The finest beef, fresh salmon and homemade pastas round out the dinner menu. It is elegant dining on fine china, crystal and silver. There is no "roughing it" in the food category at Wapiti. Diana's 11 years in the guest ranch business, coupled with her experience in catering and professional training at a culinary school, ensure guests the highest quality food presented artfully and creatively. Diana is also author of 2 cookbooks, the Dude Ranchers Cookbook and The Hearty Gourmet.

The menus Diana has shared with us are the first two meals guests enjoy upon arrival. They set the stage for the rest of the week of Hearty Gourmet dining, and I'm sure you will discover they add a new dimension to "ranch cuisine."

Breakfast Menu

ORANGES IN CINNAMON SYRUP

ASSORTMENT OF JUICES

STUFFED FRENCH TOAST WITH CARAMEL
SAUCE, CINNAMON-CREAM SYRUP AND
MAPLE, FRUIT & NUT SAUCE

SAUSAGE PATTIES

Dinner Menu

SHRIMP SPREAD

SPINACH SALAD

FOOLPROOF PRIME RIB

BRAISED NEW POTATOES WITH THYME

MUSHROOMS SUPREME

GLAZED BABY CARROTS &
SUGAR SNAP PEAS

LEMON MOUSSE WITH RASPBERRY PUREE

 Recipe Included

Oranges in Cinnamon Syrup

Light, refreshing and a touch of warmth from the cinnamon describe this luscious morning starter. It's easy to prepare, as all the preparation is done the night before. Garnish this dish with a sprig of mint or a bundle of cinnamon sticks tied together with orange peel for a different presentation.

---

4 SERVINGS

1/2 cup dry white wine
1/2 cup water
1/4 cup sugar
1 cinnamon stick, broken in half
3 large oranges, peel and with pith removed, ends trimmed, each cut into 4 rounds

Combine wine, water, sugar and cinnamon stick in small saucepan. Bring to a boil. Reduce heat; cover and simmer 8 minutes. Remove from heat.

Arrange orange rounds in a single layer in a wide shallow glass pan. Pour hot syrup and cinnamon stick over oranges. Refrigerate overnight, covered.

Vanilla Cream Stuffed French Toast

The beauty of this dish is more in the topping than the toast itself. You'll see as you read the recipe for the Maple, Fruit and Nut Sauce.

4-6 SERVINGS

8 slices sourdough bread, sliced 1/2" or less
1 (8-ounce) package cream cheese, softened
4 teaspoons vanilla extract, divided
8 eggs
1-1/2 cups milk
1 tablespoon cinnamon

Preheat griddle to 250° about 45 minutes before cooking. (If using a skillet, keep heat between low and medium low.) A low temperature is essential for cooking the toast thoroughly.

Mix cream cheese and 2 teaspoons vanilla. Spread thickly on four pieces of bread, then top with another piece of bread. Beat together eggs, cinnamon and the remaining 2 teaspoons of vanilla. Add milk and whisk together until well blended. Pour 2/3 cup of milk and egg mixture in a 9" X 13" pan or sided cookie sheet and place stuffed bread on top. Allow to soak for at least 15 minutes. Turn stuffed bread over, adding the remainder of milk and egg mixture to dish and allow to soak for another 15 minutes. Butter griddle well, cook 10-15 minutes per side, or until bread is cooked through and is medium to dark brown on surface. Serve immediately with Caramel and/or Maple, Fruit and Nut Sauce (recipe follows).

Maple, Fruit and Nut Sauce

Use this sauce liberally on the French toast as well as over pancakes, ice cream, your fingers — you get the idea. It's scrumptious and will become a family favorite in no time. I used dried cranberries and a mixed dried fruit mix. I like Diana's suggestion of fruit and would recommend her selection if you are not partial to dried apples and pears, which constitute the bulk of dried fruit mixes.

YIELD = 3 CUPS

2 cups maple syrup
1 tablespoon vanilla
1/2 cup chopped walnuts, lightly toasted *
1/2 cup sliced almonds, lightly toasted*
1 cup chopped mixed dried fruit (can use dried cherries, golden raisins, dried apricots and prunes if a mix is not available)

In a saucepan, heat maple syrup and vanilla over moderate heat, stirring until warm. Add remaining ingredients and bring just to a boil. Remove from heat and serve.

*To toast nuts, see page 11.

Mushrooms Supreme

Lush, rich, decadent. Like kissing Nicholas Cage's pouty lips. A little too much is deadly and just a taste is teasing. Not that I've actually kissed Nicolas before, heck, I've never even met the guy. But he is my favorite actor, and I have likened eating this dish to what I think it might be like to kiss him — not that I want to either (my husband is reading this book and so is my scrupulous father). I think the best approach is moderation and I think I best be quiet. Enjoy the dish.

4 SERVINGS

1 pound fresh white button mushrooms, quartered
3 tablespoons butter, melted
1 tablespoon beef bouillon granules
1/2 cup very hot water
1/4 cup butter (1/2 stick)
2 tablespoons flour
1/2 to 1 teaspoon black pepper
1/2 teaspoon salt, or more to taste
1/2 cup heavy cream
1/4 cup sour cream
Fine dry bread crumbs to cover
Grated Parmesan cheese to cover

Preheat over to 350°. Heat 3 tablespoons butter in a medium sauté pan and cook mushrooms until just giving off liquid (approximately 5 to 10 minutes). Dissolve bouillon in hot water and set aside. Melt 1/4 cup butter in large, nonstick pan. Add flour and stir until smooth. Blend in pepper, cream and bouillon broth. Stir in sour cream. Stir in cooked mushrooms. Leave in pan to bake or transfer to a baking dish. Top with breadcrumbs and cheese. Bake uncovered at 350° for 30 to 45 minutes, or until breadcrumbs are nicely toasted and mushrooms are hot.

Lemon Mousse with Raspberry Puree

Light and airy, this mousse has nothing in common with chocolate mousse, other than the name. Slightly tart and almost effervescent, this dessert is a contradiction in motion and the perfect ending for a heavy meal.

6 SERVINGS

1 envelope unflavored gelatin
2 tablespoons white wine
1/2 cup lemon juice

3 eggs, separated
1/2 cup sugar, divided
1 cup heavy cream, whipped

In double boiler, sprinkle gelatin over wine until soft. Add lemon juice. Stir over simmering water until gelatin is dissolved. Cool.

Beat egg yolks with 3 tablespoons sugar until thick, light-colored and ribbons form when you lift the spoon. Slowly add slightly cooked gelatin mixture and stir, blending thoroughly.

Beat egg whites until foamy. Add 5 tablespoons sugar gradually and beat until meringue holds soft peaks. Add whipped cream to egg yolk mixture and fold this into the meringue mixture gradually. Chill at least 2 hours.

Raspberry Puree

YIELD = 2 CUPS

1 (10-ounce) package frozen raspberries, thawed and drained, juice reserved
1 tablespoon lemon juice
2 tablespoons sugar
1 tablespoon Kirsch or cherry brandy (optional)

In blender, combine thawed and drained raspberries, lemon juice, sugar and brandy. Puree. Strain in a small-holed sieve until only seeds remain in sieve. Add enough reserved raspberry juice to thin to a usable consistency.

To serve, drizzle puree over mound of mousse on each dessert plate and garnish with fresh raspberries and sprig of mint.

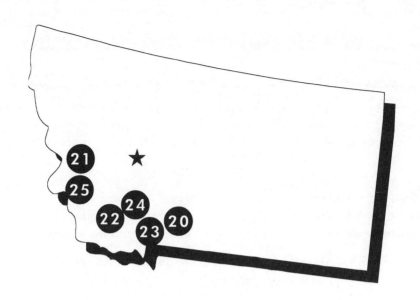

MONTANA

B Bar Guest Ranch

818 Tom Miner Creek Road
Emigrant, MT 59027
(406) 848-7523

Season: January and February
(winter)

June through September
(summer)

Capacity: 32

Accommodations: 6 cozy guest
cabins sleeping 4-5 guests each, 3
rooms in the main lodge and 1
small apartment for 3-4 guests

Activities: Winter — cross-country
skiing, snow shoeing, naturalist
guided tours, sleigh rides and hot
tub soaking. Summer — working
cattle rides, hiking, tennis, lawn
games, naturalist talks and walks,
cookouts, wagon rides,
outstanding fly-fishing and guest
rodeo

Rates: $1,950-$3,250 per person
weekly, call the ranch for details

"We salute your choice to keep this wild beauty in
its pristine form. Seldom have our days been so
full, rich, warm and deeply refreshing." This
comment from B Bar guests Maureen and Paul
Draper sums up the underlying philosophy of
ranch founders Maryanne Mott and Herman
Warsh. Maryanne, a professional photographer
and Herman, a lifelong educator, have worked
hard to maintain the raw beauty of their piece of
Paradise Valley in southwestern Montana.

Located about 20 minutes north of Yellowstone
National Park, the 20,000 acre B Bar offers
unrestricted views of the Absaroka and Gallatin
Mountain ranges as well as abundant wildlife
viewing. Large herds of elk call this section of
Montana home, and moose, eagles, and bears
share this natural habitat. Tom Miner Creek, a
feeder creek to the famous Yellowstone River,
meanders through the property and offers native
cutthroat trout. Nearby is some of the country's
best fly-fishing both in Yellowstone Park and the
famous spring creeks just south of Livingston,
Montana.

The B Bar ranch was homesteaded around the
turn of the century and has since been a working
cattle ranch. Today, Maryanne and Herman own
and manage a conservation herd of Ancient White

Park cattle, Suffolk Punch draft horses and a commercial herd of organic beef cattle. The draft horses are quite unique and these highly trained animals perform much of the work at the ranch, including grooming ski trails and logging. The ranch also boasts an enormous organic vegetable and herb garden, a high-altitude orchard and a host of other plants, shrubs and trees that dot the landscape and make the B Bar a radiant tribute to mother nature.

As a working cattle ranch, its guests have the opportunity to join in the management program and participate right alongside the ranch hands. The time of year you visit will determine the types of ranch work that will be available to you. Moving the cattle from pasture to pasture to preserve the health of the ecosystem is a summer-long activity. In the fall, there are calves to be weaned, firewood to be gathered and split and orchards to harvest. You don't have to engage in the ranch chores at all if you don't want to and there's a host of other activities to occupy your time such as hiking, fishing, pleasure riding and white-water rafting through Yankee Jim Canyon. Regardless of how you decide to spend your time, it's likely you won't forget your vacation at the B Bar or the majestic beauty of southwestern Montana.

The ranch's professional Chef Amy Jones has 9 years of experience as well as formalized training from the Culinary Institute of America, and creates fresh, high country cuisine utilizing the ranch's gardens and orchard. Chef Amy has access to over 15 different kinds of lettuces and greens from the garden and the salads she creates are always a highlight with the guests. Most meals are served buffet style and once a week Chef Amy prepares a special sit-down dinner, creating palette-pleasing and lasting memories. Sample the recipes provided by Chef Amy and see why guests continue to return to the B Bar for more.

B.Hillis

Breakfast Menu

FRESH FRUIT PLATTER

SCRAMBLED EGGS AND BACON

HOME FRIES

MARYANNE'S SOUR CREAM ✪
COFFEECAKE

Dinner Menu

CALIFORNIA NORI ROLLS ✪

B BAR SALAD WITH HERMAN'S ✪
MUSTARD VINAIGRETTE

FILET OF SALMON ✪
WITH TOMATO BEURRE BLANC

LEEKS WITH RED WINE

ROASTED GARLIC MASHED
POTATOES

POACHED PEAR TART

✪ Recipe Included

Maryanne's Sour Cream Coffeecake

Moist and tender, this coffeecake is simple to make and provides a delectable addition to your morning coffee routine. The cut cake is attractive, too, with the nutty brown filling contrasting against the cream-colored cake.

10 SERVINGS

1-1/4 cup butter (2-1/2 sticks), softened
1 cup sugar
2 eggs
1 cup sour cream
1/2 teaspoon vanilla
2 cups all-purpose flour
1 teaspoon baking soda
1 teaspoon baking powder

1 tablespoon cinnamon
1/2 cup sugar
1/2 cup chopped walnuts

Preheat oven to 350°. Grease and flour a 10" tube pan or fluted tube pan. In an electric mixer, mix butter and sugar until creamy, about 2 minutes. Add eggs, sour cream and vanilla and mix for 2 minutes. In a small bowl, sift flour, baking soda and baking powder. Add to butter mixture and mix on low speed for 2 or 3 minutes or until well-blended.

In a small bowl, mix the cinnamon, 1/2 cup sugar and chopped walnuts. Pour 1/2 batter into prepared pan. Top with 1/2 the cinnamon mixture. Cover with the remaining batter and top with the remaining cinnamon mixture. Run a knife through the cinnamon mixture all the way around the pan to cut the mixture into the batter. Bake for 30 to 40 minutes, or until a toothpick inserted in the cake comes out clean.

California Nori Roll

I generally view sushi with a wary eye, but this little tidbit is delicious and spicy. I love the kick from the wasabi (Japanese horseradish). I didn't know that sushi doesn't have to contain raw fish until I went to culinary school and learned that the correct term for the raw fish version was "sashimi" and that "sushi" is a generic term for seaweed wrapped rice and vegetables.

40-48 PIECES (1 NORI ROLL EQUALS 10-12 PIECES)

1 cup cooked sushi rice
1 tablespoon rice wine vinegar
1 tablespoon wasabi powder*
1/8 teaspoon water
4 pieces pickled ginger, thinly sliced
Cut each of the following into thin strips, 1/4" wide by 1-1/2" long (julienne):
    1/2 carrot
    1/2 red pepper
    1 scallion, green part only
    1/2 yellow squash, outer yellow part only
4 seaweed sheets (Nori)

Combine cooked sushi rice with rice wine vinegar and mix well. Combine wasabi powder and water to form a paste. Place a seaweed sheet flat on the table, long side facing you. Evenly spread 1/4 teaspoon of wasabi paste on seaweed sheet. (You can use more if you can stand the heat.) Top with 1/4 cup cooked sushi rice and spread out to all corners, leaving a 1/2" border on top. With your finger, make a groove along the center of the rice, lengthwise. Lay a thin slice of ginger, 1 strip each of carrot, red pepper, scallion and yellow squash in this groove, all the way across. You will need to use about 4 or 5 sets of this filling to go all the way across. Wet the exposed 1/2" seaweed border and roll the seaweed up from the bottom, pressing firmly so it is tight. Using a very sharp knife, moistened with warm water, cut the roll into 10 to 12 (1") pieces. Arrange on a platter, cut side up.

*Wasabi powder is available through A.J.'s Fine Foods. See "Sources" section on page 236.

B Bar Salad with
Herman's Mustard Vinaigrette

The B Bar uses organically grown lettuces and greens from its garden and greenhouse. The salads may contain a variety of greens on a given day, but will likely contain at least one of the following lettuces: Romaine, Green or Read Leaf and Butter lettuces. Also thrown in might be Arugula, Cilantro, Swiss Chard, Mizuna, Radicchio, Watercress, Mustard greens, Sorrel, Papalo, Kale or Collard greens. Most salads are garnished with edible flowers.

Use about 1-1/2 cups of lettuce/greens per person and 1-1/2 tablespoons dressing per salad. For 4 people, use 6 cups of lettuce/greens and 1/3 cup dressing, tossing well to coat. Throw in shredded carrots, thinly sliced red onion and fresh sliced cucumber for a perfect salad mix. The vinaigrette tastes better if you let the flavors blend for at least 4 hours or overnight before you serve it.

Herman's Mustard Vinaigrette

YIELD = 1-1/2 CUPS

1 teaspoon chopped fresh tarragon (or 1 tablespoon dried)
1/4 cup Dijon-style mustard
1 cup olive oil
1/4 cup white wine vinegar
1 teaspoon sugar
Black ground pepper to taste

Mix all ingredients together in a blender or in a medium bowl with a whisk. Taste to see if it needs a pinch of salt. Store in the refrigerator for at least 4 hours or overnight. Will keep in the refrigerator 4 days.

Filet of Salmon
with Tomato Beurre Blanc

Beurre Blanc is the French name for a white wine butter sauce. Many Chefs today are adding additional ingredients to give a new twist on the old classic. I've seen Lime Beurre Blanc and Mango Beurre Blanc and a host of others. This Tomato Beurre Blanc is light and tangy and is a perfect accompaniment to the broiled salmon. You can grill the salmon if you prefer.

2 SERVINGS

2 (8-ounce) salmon filets
2/3 cup sour cream, divided
2 tablespoons Dijon-style mustard
1 pound (about 4 large) tomatoes, peeled, seeded and chopped
2 teaspoons olive oil
Salt and pepper to taste

1/8 teaspoon fresh thyme leaves (or 1/2 teaspoon dried)
1 tablespoon minced shallot
2 tablespoons white wine
2 tablespoons white wine vinegar
1/2 cup butter (1 stick), cut into 8 pieces and very cold

Mix only 1/2 cup sour cream and mustard together and coat salmon on both sides. Set aside.

Preheat broiler. Heat oil in a skillet over medium heat and add tomatoes, cooking for about 10 to 15 minutes, until juices are thick and somewhat reduced. Stir in thyme and season with salt and pepper. In a separate skillet over medium heat, combine shallot, white wine and vinegar. Cook until the mixture is almost evaporated, with only about 1 tablespoon left. Whisk in remaining sour cream and heat through, about 2 minutes. Turn heat to the lowest setting. While swirling pan, add butter, 1 piece at a time. When the piece is almost completely melted, add the next piece of butter and continue to swirl the pan. After 3 or 4 pieces have been added, remove pan from heat and continue to swirl and add butter. After the last piece has been added and melted, pour butter mixture into the warm tomato mixture. Season with more salt and pepper if necessary. Keep warm while broiling the salmon, but watch the heat. Too much heat will "break" the sauce, and it will separate. Broil the salmon until opaque, about 3 to 5 minutes per side. Spoon 2 to 3 tablespoons sauce on a plate and top with salmon. Garnish with fresh thyme sprig and/or lemon wedge.

Bear Creek Lodge

1184 Bear Creek Trail
Victor, MT 59875
(406) 642-3750
Website: http://www.bear-creek-lodge.com
E-mail: info@bear-creek-lodge.com

Season: Year-round

Capacity: 16

Accommodations: 8 luxurious guest rooms, each with 2 double or queen beds and a private bath. Common areas include living room with fireplace and overstuffed chairs, dining room, conference room (with communication equipment), library with billiards table, exercise room with sauna and hot tub.

Activities: exceptional fly-fishing; hiking; horseback riding; white water and scenic rafting; cultural and historic sites; wildlife refuges; and golf and tennis nearby

Rates: $150-$210 single/$180-$300 double occupancy per night

Bear Creek Lodge is not a guest ranch. But because of its location, the activities, the gourmet food and the humble business philosophy of the owners, it is included in this book because the only thing it's <u>not</u> is a ranch. Nestled at the edge of the Selway-Bitterroot Wilderness area, Bear Creek Lodge is the perfect getaway to experience the great outdoors in luxurious comfort. Fly into Missoula, Montana, and in roughly 45 minutes you'll be crossing over Bear Creek through a covered wooden bridge, driving past a pastoral barn and open pasture. Follow the wooden boardwalk through the trees to the sudden grandeur of Bear Creek Lodge. The lodge itself is built from massive logs harvested near Yellowstone Park after the disastrous fires of 1988. Each of the spacious guest rooms conjures up images of a tamed West, with antiques, hand-woven blankets and just the right amount of Western decorative touches. Welcome to the Turneys' piece of paradise.

Most of Roland and Elizabeth's guests come to the Bitterroot because of their love of nature. With private access to Bear Creek, anglers enjoy the wild, native trout. Nearby are blue ribbon trout rivers such as the Bitterroot, Big Hole and Blackfoot. The anglers and non-anglers alike cherish float trips on the Big Hole River. Don't miss this scenic journey down one of the most beautiful rivers in the West. If horseback riding is more your fancy, then set out on one of the many trips up Bear Creek Canyon with an experienced backcountry guide. Hikers have so many options, making a choice is difficult, but always rewarding. Day hike into the mountains along boulder-strewn Bear Creek as it meanders through the canyon. Or plan a serious hike in the Bitterroot Wilderness, and look for white-tailed deer or elk or moose splashing in small ponds.

Within an hour's drive, you can visit several historic sites, including Saint Mary's Mission, the first permanent white settlement in Montana, or the Daly Mansion, built in the late 1880's as a summer home for one of the state's copper barons. The Big Hole Battlefield, where the Nez Perce Indians and U.S. Army clashed in 1877, is a must-see for history buffs. For nature lovers, the National Bison Range, one of the last remaining herds of American Buffalo, is just north of Missoula, or visit the Lee Metcalf and Otto Teller wildlife refuges. There are so many things to do and see while staying at Bear Creek Lodge, you'll need to extend your vacation so that you can spend some time just relaxing in the cozy and comfortable living room or curl up with a good book in the lodge's library.

Roland and Elizabeth are also prepared to welcome business groups with a conference room and visual aids and communication equipment. Being life-long educators, your hosts understand the importance of environment and atmosphere for productive business meetings. Use the lodge as your workshop facility or business retreat and combine fruitful meetings with team-building outdoor activities.

Bear Creek Lodge has received national attention for its ideal location, generous hospitality, comfortable surroundings but most of all for its stellar food. Elizabeth is the quintessential gourmet cook. "Country Inn Cooking with Gail Greco" featured Elizabeth and her cooking in a recent PBS show. Guests are always thrilled and surprised at the quality and presentation of the daily meals at Bear Creek Lodge. Elizabeth loves cooking and it shows as she puts her heart and soul into fresh baked breads, homemade soups and multi-cultural-infused main entrees. Prior to guests' arrival, Elizabeth has already discussed their diets and culinary likes and dislikes. She is also focused on providing healthful, balanced meals and many of her dishes are low in fat and calories but big in flavor. The presentation of the food is absolutely gorgeous, though if you asked Elizabeth, she would say the presentation is secondary to the taste and flavor of her meals. Most of Elizabeth's creations contain herbs from her garden and the freshest local organically grown fruits and vegetables. Her cuisine is unique and her presentation is artful. Enjoy these treasured favorites Elizabeth has graciously shared with us.

B.Hillis

Breakfast Menu

ORANGE SLICES
WITH STRAWBERRIES

CHILI EGG BAKE
WITH TOMATO SALSA

BEAR CREEK GRANOLA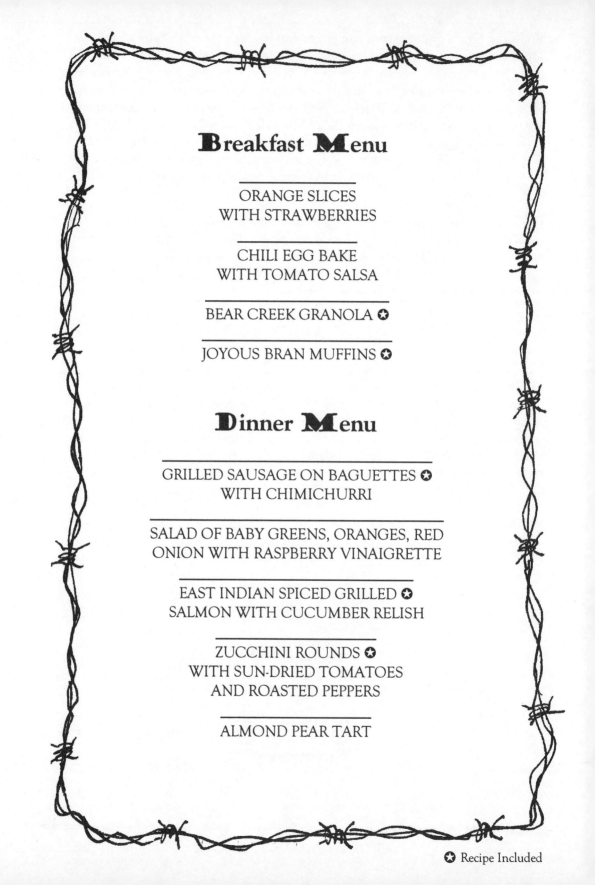

JOYOUS BRAN MUFFINS ✪

Dinner Menu

GRILLED SAUSAGE ON BAGUETTES ✪
WITH CHIMICHURRI

SALAD OF BABY GREENS, ORANGES, RED
ONION WITH RASPBERRY VINAIGRETTE

EAST INDIAN SPICED GRILLED ✪
SALMON WITH CUCUMBER RELISH

ZUCCHINI ROUNDS ✪
WITH SUN-DRIED TOMATOES
AND ROASTED PEPPERS

ALMOND PEAR TART

✪ Recipe Included

Grapevine Canyon Ranch, Arizona (above and page 22). Rancho de los Caballeros, Arizona (below and page 26).

Tanque Verde Ranch, Arizona (above and page 35). White Stallion Ranch, Arizona (below and page 44).

Alisal Guest Ranch and Resort, California (above and page 48). C Lazy U Ranch, Colorado (below and page 71).

King Mountain Ranch, Colorado (above and page 93). Vista Verde Ranch, Colorado (below and page 124).

Wit's End Guest Ranch, Colorado (above and page 134). Bear Creek Lodge, Montana (below and page 158).

Craig Fellin's Big Hole River Outfitters, Montana (above and page 167). Complete Fly Fisher, Montana (below and page 179).

Cibolo Creek Ranch, Texas (above and page 197). Flying A Ranch, Wyoming (below and page 214).

T Cross Ranch, Wyoming (above and page 221). UXU Ranch, Wyoming (below and page 229).

Bear Creek Granola

This crunchy granola has a terrific blended flavor of nuts and fruit. It also seems lighter and more natural than some granolas and I think it's because there is no oil and it relies on the intrinsic properties of honey and apple juice to bind it together. If you use sweetened coconut, the granola will be very sweet and have a predominantly coconut flavor, which is not necessarily a bad thing. Elizabeth suggests serving this with a combination of bananas and non-fat yogurt that have been blended together.

YIELD = 9 CUPS

4-1/2 cups old-fashioned rolled oats
2 cups coconut (preferably unsweetened)
1-1/2 cups slivered or chopped nuts —
　　almonds, walnuts, pecans, hazelnuts, and/or whole sunflower seeds
1/4 cup honey
3/4 cup apple juice concentrate
1-1/4 cups dried fruit —
　　raisins, cranberries, dates, and/or other fruit, chopped to raisin-size

Preheat oven to 350°. Toss the oats, coconut, and nuts together in a large baking pan with 2"+ sides. Put honey and concentrate in a small, microwave bowl, heat one minute to thin honey and whisk. Pour over the oat mixture, tossing until coated. Bake, stirring every 8 to 10 minutes, until oats are golden. (It took me about 30 minutes in total). Remove and continue stirring at same interval until cool. Stir in the dried fruit. When completely cool, store in a sealed plastic container.

Joyous Bran Muffins

These muffins have a dense, heavy texture and are surprisingly moist. The flavor is intense and dark and really good despite the absence of fat content. I prefer to make the jumbo muffins, which brings the yield down to 10 to 12 muffins. A fringe benefit is the batter will keep in the refrigerator in an airtight container for up to 2 months! Yep! You can make a big batch of batter (double this recipe) and have fresh baked muffins every morning for weeks! Make sure there is no air in your airtight container or the batter will ruin.

YIELD = 20-24 MUFFINS

1 cup ALL-BRAN®
1 cup raisins
1 cup boiling water
1/2 cup sugar
1/4 cup fruit puree*
1 cup molasses

2 eggs, beaten
2 cups buttermilk**
1-1/2 cups dry ALL-BRAN®
2-1/2 cups all-purpose flour
2-1/2 teaspoons baking soda
1/2 teaspoon salt

If cooking now, preheat oven to 400° and grease muffin tins. In a small bowl, pour water over the 1-cup of ALL-BRAN® and raisins and let sit until cool. Put sugar in large (really large as this batter expands) bowl, then add the following, in this order, stirring as you go: fruit puree, molasses, eggs, buttermilk, remaining 1-1/2 cups of ALL-BRAN®. In a medium bowl, combine flour, baking soda, and salt then add to the sugar mixture. Stir in soaked ALL-BRAN® and raisins. Fill muffin cups 3/4 full, cook in preheated 400° oven for 20 minutes. Cool slightly before removing from pan.

*There is new a commercial baking product called lighter bake that is a substitute for oil and butter. You may also make your own by stewing together a combination of prunes and apples and puree or try just plain applesauce.

**2 cups skim milk with two tablespoons of vinegar left to sit a few minutes makes a fine substitute.

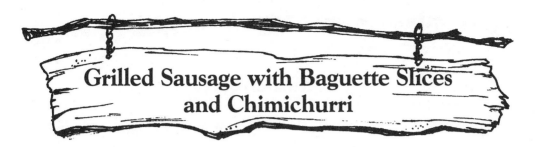

Grilled Sausage with Baguette Slices and Chimichurri

An extraordinary appetizer both in taste and appearance. If you make the sauce a few hours in advance, there is not much work to putting these fabulous little tidbits together. Elizabeth recommends turkey sausage, but I used a smoked beef sausage. It was so delicious I could have eaten just this for dinner. Use any kind of baguette-shaped bread, such as French, Italian or sourdough. Elizabeth says that it does make a great lunch treat if you put the grilled sausage and Chimichurri on a 3" piece of warmed bread and serve with a salad.

---

8 APPETIZER SERVINGS

4 large sausages (about 1-1/2 pounds), preferably turkey
16 slices bread, about 3/4 inch thick from warmed loaf
1 batch Chimichurri sauce (recipe follows)

Preheat grill to medium high. Grill sausage until cooked through and slightly crisp. Slice diagonally. Spoon 1/2 to 1 teaspoon of sauce on each slice of bread and top with one or two pieces of sausage, depending on size and shape. Serve with more sauce and napkins.

B.Hillis

Chimichurri

Though not particularly hot by my standards, Elizabeth says this is one of the spicier condiments in Argentina, a country not really known for hot foods. This relish-like sauce is incredibly flavorful. Combined with sausage and bread, it makes a unique appetizer that will have your guests guessing which herbs and spices it contains. This will keep well in the refrigerator, although the parsley will lose its vibrant green color. Serve this with grilled beef, and lamb as well as Elizabeth's favorite: Grilled Sausage with Baguette Slices and Chimichurri.

YIELD = 1 CUP

1/2 large onion, rough chopped
2 cloves garlic
1/2 jalapeno, seeded
1/2 bunch of parsley, tops only
1 tablespoon fresh oregano
1-1/2 teaspoons salt
1 teaspoon freshly ground black pepper
1/2 cup olive oil
1/4 cup balsamic vinegar

In the bowl of processor with chopping blade, place onion, garlic and jalapeno. Pulse a few times, and then add parsley, oregano and pulse until mixture is finely chopped. Remove to bowl, then add vinegar, salt, and pepper to processor and drizzle in oil as it is mixing. Mix thoroughly, then pour over herb mix and combine. Let stand at room temperature at least 2 hours before serving.

East Indian-Spiced Grilled Salmon

Exotic flavors coat your palette with every bite of this juicy, fleshy fish. It's an unusual flavor experience and the cucumber relish accompaniment is just the perfect addition to this light and luscious treat. It makes me want to try other non-curry type items from India.

6-8 SERVINGS

Marinade:
2 tablespoons olive oil
2 tablespoons fresh orange juice
3 tablespoons grated gingerroot
1 tablespoon cumin
1 tablespoon ground coriander

1 teaspoon turmeric
1 teaspoon salt
1/2 teaspoon cayenne
3 pounds (6-8 ounces portions) salmon
    filets or steaks

Combine marinade ingredients in a small bowl. About half an hour before grilling, brush both sides of salmon with marinade, cover and let sit at room temperature. Meanwhile, preheat grill and coat grill with cooking oil. When fire is medium hot, cook about 5 minutes, then flip. Filets will take 8 to 10 minutes total cooking time, steaks slightly longer. Fish is cooked when it is firm to the touch, (with steaks check both sides for doneness). Remove immediately from grill and serve with Cucumber Relish (recipe follows). If fish is left on a grill to cool, it will surely stick, so remove to platter immediately.

Cucumber Relish

An English cucumber is similar to a "hothouse" cucumber and has fewer seeds and a slightly thinner skin than the common grocery store variety. I think it can be less bitter, too. This is a pretty relish and looks great on the grilled salmon dish, and its taste is refreshing.

6-8 SERVINGS

1/2 cup seasoned rice vinegar
1 teaspoon sesame oil
1 tablespoon minced fresh ginger

1/2 teaspoon salt
1/2 English cucumber, quartered
    lengthwise and cut in 1/4 inch slices

Whisk together the vinegar and oil. Add ginger and salt to the oil mixture and pour over cucumber. Let relish sit at room temperature at least one hour. May be prepared up to 12 hours ahead. Serve with East Indian Spiced Grilled Salmon.

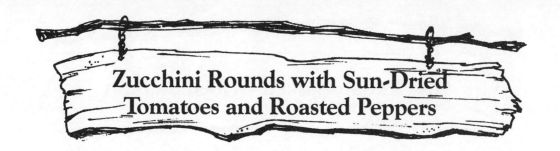

Zucchini Rounds with Sun-Dried Tomatoes and Roasted Peppers

These are fun to make and fun to eat! They look quite fancy and taste exquisite. The combination of blue cheese and sun-dried tomatoes is fantastic. Do them up in advance and use them as appetizers as an alternative.

8 SERVINGS

6 small zucchinis
1/2 cup drained and squeezed sun-dried tomatoes in oil,
1/4 cup roasted red pepper or 1 red pepper, roasted and peeled
1 tablespoon chopped fresh basil leaves
1/8 cup blue cheese, crumbled and loosely packed
3 tablespoons freshly grated Parmesan
1/2 teaspoon freshly ground pepper
Non-stick cooking spray

Preheat oven to 400°. Slice the zucchini into 1/2 inch rounds and remove most of the pulp with a melon baller (Elizabeth saves the pulp for the soup pot.) Put tomatoes in the bowl of the processor with blade, pulse to chop, then add peppers and basil. Pulse again but only chop, don't puree. Add cheese and pulse just to blend. Season with salt and pepper if desired. Lightly spray a cookie sheet. Fill zucchini rounds and place on cookie sheet. Bake 8 to 10 minutes and serve.

Craig Fellin's Big Hole River Outfitters

P.O. Box 156
Wise River, MT 59762
(406) 832-3252
E-mail: wsr3252@montana.com

Season: mid-June through early October

Capacity: 8-10

Accommodations: Three cozy log and cedar cabins named Pattengail, Big Hole and Red Rock

Activities: Trophy fly-fishing and more trophy fly-fishing; historic sightseeing; Signature Jack Nicklaus golf course nearby; pitching green (actually a par 3 hole) on the lodge premises

Rates: $2,150 per person (6 nights/5 days, double occupancy), includes all meals and lodging and 1 guide per 2 fishers

Big Hole River Outfitters is for serious fly-fishers and not necessarily only experienced fishers because the operation offers quality teaching/ guiding for all levels of expertise. Even if you've never pick up a fly rod before, you can walk away a week later and have a good foundation of angling skills. Craig Fellin provides first class accommodations and amenities, professional, knowledgeable guides, and a talented Chef to prepare gourmet food that will turn your trip into an odyssey. Mother Nature provides the scenery and the fish.

The lodge is situated in southwestern Montana, just 50 minutes south of Butte. Surrounded by wilderness and straddling the Wise River, Craig's property boasts rustic and cozy log cabins. A gorgeous river rock and lodgepole pine main lodge supply a sprawling, scenic view of the Wise River Valley and Pioneer Mountains. Open your window at night early in the season and you can hear the river traversing through the meadow. Early risers are treated to wild elk grazing on the dew-kissed grassy meadow a short walk from the lodge. And the whole area is prime moose habitat.

Although many non-anglers sojourn to Craig's operation and find plenty to occupy their time, Big Hole River Outfitter's is a fly-fisher's dream. The week is filled with a variety of fishing excursions, some wading, some floating, on famous water such as the Big Hole River, one of Montana's finest blue ribbon rivers. You might also catch the fabled Beaverhead River or discover the challenges of crystal-clear spring creeks. Craig's access to a private 20-acre spring lake brings you nose to nose with trophy trout up to 24-inches. When I asked Craig what was the one thing I shouldn't miss, he said "the first 20-inch trout that comes up to your fly." I've discovered he has a sense of humor, but I don't think he was joking.

I just know I'm going to embarrass him by saying this, but Craig is one of the most kind, humble and gracious men I know (and handsome, too!). As one of his guests, you know that you will be treated with respect and kindness and you won't want to leave his little corner of paradise. His own philosophy of host hospitality permeates throughout his staff. With only 8 to 10 guests a week, the intimate feeling is reiterated daily by a courteous, gracious and attentive staff. And the food? That's something to write home about.

Before you arrive, Craig has sent you a questionnaire that captures not only your activity interests, but also your food preferences. His Chef wants to know what you like and especially what you don't like. Lanette Evener has been at the lodge for four seasons and exudes the same humbleness as her boss. She says she learned a tremendous amount from the previous Chef and credits her passionate love for cooking as another reason for her success. During the week you might have grilled salmon for dinner, with a mustard dill sauce, paired with a colorful and tasty rice pilaf and finish with a fresh baked almond cake and homemade ice cream. Lunch could be marinated tri-tip roast on a fresh baked roll and Mexican corn salad. Munch on sweet and tart lemon bars to tide you over until dinner. The food is fabulous and you'll remember it as much as you do the big fish you caught and the one you didn't.

Breakfast Menu

FRESH ORANGE JUICE

SEASONAL FRUIT PLATE

OATMEAL PANCAKES WITH ✪
CARAMELIZED APPLES AND
TOASTED WALNUTS

CRISPY BACON

Dinner Menu

PEAR STILTON
BLUE CHEESE SALAD

ROSEMARY GINGER ✪
ROASTED CHICKEN

CHEESY GARLIC MASHED ✪
POTATOES

STEAMED BROCCOLI

FRESH PEACH TART

✪ Recipe Included

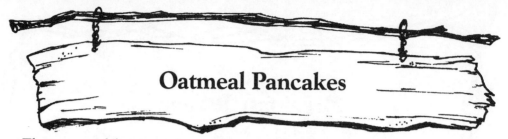

Oatmeal Pancakes

These are so delicious and taste like oatmeal cookies sans raisins. The apples are warm and spicy and complement the pancakes nicely. I love walnuts and their crunch adds just the right touch. The recipe makes 12 (4-1/2") pancakes which is 4 servings unless, of course, you're feeding my husband, then it's only 2 servings. For a very slim man he can pack away these pancakes. The good news is I didn't have to feed him again until dinner.

4 SERVINGS

2 cups oats
2 cups buttermilk
1/2 cup flour
4 tablespoons brown sugar
1 teaspoon baking soda

1 teaspoon baking powder
1 teaspoon cinnamon
1/4 teaspoon nutmeg
2 eggs, lightly beaten
1/4 cup melted butter

In a large mixing bowl, stir the oats and buttermilk together. Let rest for 30 minutes.

In a small mixing bowl, mix the flour, brown sugar, baking soda, baking powder, cinnamon and nutmeg together. After oatmeal mixture has sat 30 minutes, add flour mixture, eggs and melted butter to oatmeal mixture. Stir just to combine. Heat a skillet or griddle to medium heat and pour a scant 1/4 cup batter for each pancake. Cook 3 to 4 minutes, until bubbles form on top and edges start to dry. (These pancakes take a little longer to cook than regular ones.) Flip cake over and cook another 3 to 4 minutes. Serve with Caramelized Apples (recipe follows), toasted walnuts and a splash of maple syrup.

Caramelized Apples:
2 apples, peeled, cored and sliced 1/4"
    thick
4 tablespoons butter

4-5 tablespoons sugar
1/4 teaspoon cinnamon
1/2 cup walnuts, toasted*

Melt butter in a medium skillet over medium heat. Add apple slices, sugar and cinnamon. Cook until apples soften slightly, 10 to 15 minutes. Place pancakes on a warm plate, top with a few apple slices and sprinkle with toasted walnuts.

*To toast nuts, see page 11.

Rosemary Ginger Roasted Chicken

The aroma of this chicken should be bottled and sold for perfume! It will make you swoon in sheer ecstasy. There is a subtle undertone of lemon and a peppery bite that's pleasantly surprising. The skin is crispy brown and the chicken just melts in your mouth. If there are any leftovers, make a flavorful chicken salad.

4 SERVINGS

1 whole chicken (2 to 2-1/2 pounds)

5 tablespoons olive oil
3 tablespoons minced garlic
1 tablespoon black pepper
2 tablespoons chopped fresh rosemary
1 tablespoon chopped fresh parsley
1 tablespoon chopped fresh sage
1 tablespoon chopped fresh thyme
3 teaspoons salt
1 teaspoon cayenne
1 teaspoon chopped fresh ginger
1 teaspoon grated lemon peel
Juice of 1/2 lemon

Discard any contents inside the chicken and rinse the bird thoroughly. Cut the backbone out, turn the bird over and break the ribs by pushing down hard with your palm. This makes the bird lay fairly flat.

Combine all other ingredients in a medium bowl. Pull the flesh away from the bird and using your hands, rub the marinade into the meat, going all the way to the legs and thighs, as well as the breasts. Rub remaining marinade on the outside skin. Cover and refrigerate for 2 hours.

Preheat oven to 425°. Place chicken on a rack in a roasting pan. Fill the bottom of the pan with 1/4" hot water. Roast for 40 minutes or until done. Let it rest for 5 minutes, then cut into quarters or remove all the meat from the bones and transfer to a warm platter for serving.

Cheesy Garlic Mashed Potatoes

Do you like cheddar cheese on your baked potato? If you do, you'll love these mashers. The cheddar gives them a deeper flavor dimension than plain mashed potatoes. Leaving some of the potato skin gives a good contrast in texture. I like to sprinkle a little bit of cheddar on top right before serving.

4 SERVINGS

4 baking potatoes (about 2 pounds)
1 head garlic
1 teaspoon olive oil
1/4 cup sour cream
1/4 cup + heavy cream
3 tablespoons butter
1/2 cup shredded cheddar cheese plus more for garnishing
Salt and pepper to taste

Peel half the potatoes and rinse (both peeled and unpeeled). Cut into quarters and place in a deep saucepan and cover with water. Boil 30 to 40 minutes or until very tender.

Meanwhile, preheat oven to 400°. Cut off just enough of the top of the garlic to expose the cloves. Drizzle with the olive oil and wrap in aluminum foil. Bake 35 to 45 minutes or until cloves are soft. Remove from oven and cool slightly. Remove cloves from skin and mash into a paste. Set aside.

Drain potatoes and return to same cooking pan. Add sour cream, heavy cream, butter, cheese, garlic and salt and pepper. Mash with potato masher or if you prefer whipped potatoes, beat with an electric mixer until smooth. Garnish with more cheddar cheese if desired.

Lone Mountain Ranch

Lone Mountain Ranch

P.O. Box 160069
Big Sky, MT 59716
(406) 995-4644
Website: http://
    www.lonemountainranch.com
E-mail:
    lmr@lonemountainranch.com

Season: June through October;
December through mid-April

Capacity: 80-90

Accommodations: Varies to
meet guest needs, from one-
room log cabin to larger log
cabins sleeping 10. Six rooms
in ridge top lodge with
luxurious furnishings and a
fantastic view

Activities: Naturalist programs;
horseback riding; Yellowstone
National Park tours; nature
trail; photographer tours; fly-
fishing trips; climbing wall;
extensive children's program;
winter cross-country skiing;
snow-shoeing

Rates: Summer — $1,550 per
person per week, double
occupancy

Winter — $1,200 per person
per week, double occupancy

Lone Mountain Ranch is famous for its
naturalist programs and the staff takes great
pride in educating its guests about the
Yellowstone Ecological System. Located close
to both Yellowstone National Park and the Big
Sky ski area, Lone Mountain makes a great
vacation destination summer or winter. In
addition to the naturalist programs, Lone
Mountain offers a host of other activities for
the outdoor enthusiast, including an excellent
riding program and world-class fly-fishing. Of
course, it's a great place to do nothing, too,
and you can set your own schedule. If you want
to venture from the ranch, located nearby are
the Lewis & Clark Caverns, the Museum of the
Rockies, the Grizzly Discovery Center and some
of the most famous trout fishing rivers in the
U.S.

Tucked away in a private valley carved by the
North Fork, a headwater mountain stream for the
beautiful Gallatin river, the main lodge is an
impressive log and stone structure that is inviting
and cozy despite its massive exterior. Perched on a
hill, the lodge offers panoramic views that will take
your breath away. The ranch boasts a guest repeat
rate in excess of 70%, and credits its staff for
bringing guests back year after year. Most guest
comments focus on the staff and the quality of the
service.

That same warm, friendly service extends into the
food operation at Lone Mountain as well, the
direct result of Chef Bob Hayes and his customer
satisfaction philosophy. Chef Hayes manages a

staff of 16, ranging from Culinary Institute of America (CIA) graduates to fly-fishing and skiing enthusiasts who also love to cook. They all love to work and play in the bountiful mountains of Montana. Guests are often surprised at the quality and variety of the cuisine at Lone Mountain, but Chef Hayes expects nothing less than perfection from his diverse culinary team. Judging from guests' comments, he easily achieves that goal. A recent guest stated that the food at Lone Mountain was "as good as some of the best restaurants..." they had visited. Lone Mountain has a state of the art smoker that is frequently fired up to deliver smoked crown pork, baby back ribs and smoked trout that is "out of this world." The menus change daily and always offer succulent choices featuring the freshest ingredients. Breakfast is an enormous buffet and several items rotate so those guests staying a week are not bored with the selection. Dinner is a fabulous affair and the entrée choices are what you would expect from a fine dining establishment in a large metropolitan area. Chef Hayes has graciously shared some of his most treasured recipes with us and encourages you to try them at home.

B. Hillis

Breakfast Menu

BUFFET

Or

SOUTHWEST EGGS BENEDICT
WITH CHORIZO SAUSAGE AND
JALAPENO HOLLANDAISE

BANANA CHOCOLATE CHIP ✪
MUFFINS

Dinner Menu

TOMATO HERB SOUP ✪

BASIL CRUSTED HALIBUT

LEMON ASPARAGUS

CARROT MOUSSE

POTATOES GRATIN

DERBY PIE ✪

✪ Recipe Included

Banana Chocolate Chip Muffins

An interesting combination, these muffins taste great. I love bananas and I love chocolate. I just never thought about putting the two together. Now I think about it all the time. Yum!

YIELD = 18-20 MUFFINS

1 cup sugar
1/2 cup butter (1 stick) softened
1-1/2 teaspoons baking soda
1/2 teaspoon salt
2-1/4 cups all-purpose flour
2 ripe bananas
1/3 cup buttermilk
1 teaspoon vanilla
3/4 cup semi-sweet chocolate morsels
2 large eggs

Preheat oven to 350°. In a mixer, mix sugar and butter together on medium speed. Add baking soda, salt and flour and mix again on low for 1 minute. Next add bananas, buttermilk, vanilla, chocolate chips and eggs and mix on low for 2 minutes. Fill a greased or lined muffin pan 3/4 full. Bake for 20 to 25 minutes. Remove from oven and cool 2 minutes. Remove from muffin pan and cool on a wire rack.

Tomato Herb Soup

The riper the Romas, the more flavorful this soup. The simmering tomatoes and herbs smell divine and are outdone only by the taste of this fabulous comfort food. Serve with a crisp green salad and crusty bread for a light but complete meal.

4 SERVINGS (ABOUT 5 CUPS)

15 to 20 Roma tomatoes (about 2-1/2 pounds), cored and chopped
1 large yellow onion, chopped
1 tablespoon minced garlic
3 tablespoons olive oil
1 quart chicken stock*
3 bay leaves
2 tablespoons finely chopped fresh thyme
2 tablespoons finely chopped fresh basil
2 tablespoons finely chopped fresh chives
Salt and pepper to taste

In a stockpot, cook tomatoes, onion and garlic in olive oil over medium heat until onions are translucent, about 5 minutes. Add the chicken stock and bay leaves and simmer for 15 to 20 minutes. Remove the bay leaves. Working in small batches, puree the soup and return to stockpot. (Careful! Hot liquid in a blender will shoot up fast and hard when you first turn it on. Hold the lid down very tight.) Add fresh herbs to the soup and season with salt and pepper. Simmer for 10 to 15 minutes and serve. Garnish with a teaspoon of grated Parmesan if desired.

Concentrated chicken stock is available through More Than Gourmet. See "Sources" section on page 236 for details. One (1.0-ounce) tin of Fond de Poulet Gold® makes 5 cups (you need 4 cups for this recipe).

Derby Pie

I love this crust. I make extra and freeze it. It's easy to work with and tastes so buttery when baked. As for the pie, I could get in trouble if left alone in a room with it. Decadent, rich, chocolate — need I say more?

YIELD = 1 PIE, 8 SERVINGS

Crust:
1-1/2 cups all-purpose flour
1/2 cup butter (1 stick)
1/4 to 1/3 cup milk

Filling:
1 cup sugar
1/2 cup butter (1 stick), cold and cut into chunks
2 eggs
2 generous tablespoons Jack Daniel's or other bourbon
1/2 cup all-purpose flour
1 cup cashews
1 cup semi-sweet chocolate morsels

For the crust, cut the butter into the flour with a pastry cutter or with your hands. When the mixture resembles coarse meal, slowly add the milk, stirring until incorporated. Knead the dough only a couple of times and then wrap it in plastic wrap and put it in the refrigerator for 30 minutes. Roll dough out on a lightly floured surface just a little larger than a 9" pie pan. Place in pan, push down and finish edge (like crimping with a fork).

Preheat the oven to 350°.

For the filling, beat the sugar and butter until creamy and smooth. Add the eggs, bourbon and flour and mix well. Fold in the nuts and morsels. Pour filling into unbaked pie shell and bake for 30 to 40 minutes or until firm to the touch and crust is golden brown.

The Complete
Fly Fisher

P.O. Box 127
Wise River, MT 59762
(406) 832-3175
Website: http://
   www.completeflyfisher.com
E-mail: comfly@montana.com

Season: June-October

Capacity: 14

Accommodations: Private,
spacious riverside cottages, main
lodge also on the river

Activities: fly-fishing; nature tours;
horseback riding

Rates: $2,300 per person 6
nights/5 days, includes lodging,
all meals, transportation and a
guide for every 2 anglers

Dave Decker is a very serious man. He's serious
about providing the kind of service his guests want
and deserve. The Complete Fly Fisher is an
exclusive fly-fishing resort on the banks of the
famous Big Hole River in southwestern Montana.
The operation may be small, taking only 14 guests
a week, but the service is grand. Amazingly,
women guests have influenced many of the
touches at the lodge, especially the menu and the
accommodations. Dave pays attention to what they
say, what they want, and as a result, you'll find
some positive reflections at a lodge that used to
attract predominantly men. Today, over half the
guests are women. Dave and his outstanding crew
teach fly-fishing from the bottom up. Couple that
with the breathtaking scenery, the spacious,
comfortable, riverside cottages and first-class
dining and you've got a vacation to remember.

The fishing is as challenging as you could ask for,
or as easy as you want. There are a variety of
waters around the lodge that will suit any and all
of your moods. You can find a variety of angling
situations on the Big Hole River alone, not to
mention the Beaverhead River or the lodge's
access to private water on the Wise River. If you
are new to fly-fishing, don't worry about a thing.
The Complete Fly Fisher has competent, seasoned
professional guides who love to share their
knowledge. The guides have years of experience
teaching novices and offer refresher courses for the

experienced angler. The lodge even has a fully stocked fly shop to accommodate any of your equipment needs. If fishing the entire time is not exactly what you had in mind, let the lodge arrange for scenic horseback riding with one of three different local outfitters. Just let the lodge know before you get there so space can be reserved. Nature tours and hiking excursions can also be pre-arranged.

Check out the web page (www.completeflyfisher.com). It's an extensive Website, packed with information and pictures.

While Dave and his guides take care of your angling needs, Chef Mary Creagh takes care of your nourishment. Meals are served at angler's hours (which means early in the morning and a little later in the evening) on the main lodge porch overlooking the Big Hole River. You get gorgeous views and handsome plates of fine western cuisine at the same time. Chef Creagh has been with the lodge for more than 3 years and has many years' experience as a guest ranch Chef. In the off-season, she and her husband travel to New Zealand where she is a Chef at another famous lodge. Both lodges benefit from her experience. She can truly say her cuisine is a fusion of international influence.

Breakfasts are likely to feature Smoked Salmon Omelets, Pumpkin Pancakes or Belgian Ale Waffles accompanied by an assortment of smoked sausages and bacon, homemade breads and pastries. Dinners are served by candlelight with a fine wine, selected to enhance the specific entrée. You might be treated to Bourbon-Molasses Grilled Quail with a Southern Spoonbread, or a Maine Lobster Salad on bitter greens with Sun-dried Cherry Vinaigrette. Chef Creah also accommodates vegetarians, children or anyone with a specific dietary need. The service and presentation will make you think you are in a major metropolitan star restaurant, but then you take a deep breath of clean mountain air and realize how far away you are and how glad you are to be in Big Sky country.

Chef Creah has shared with us two delightful menus and some enchanting recipes that go with them. One taste and you'll be on the phone, booking a trip to The Complete Fly Fisher.

Breakfast Menu

MONTANA BLEND COFFEE AND
ASSORTED TEAS

HOMEMADE HONEYED-MUESLI
WITH SEASONAL BERRIES, YOGURT
OR CREAM

PUMPKIN PANCAKES WITH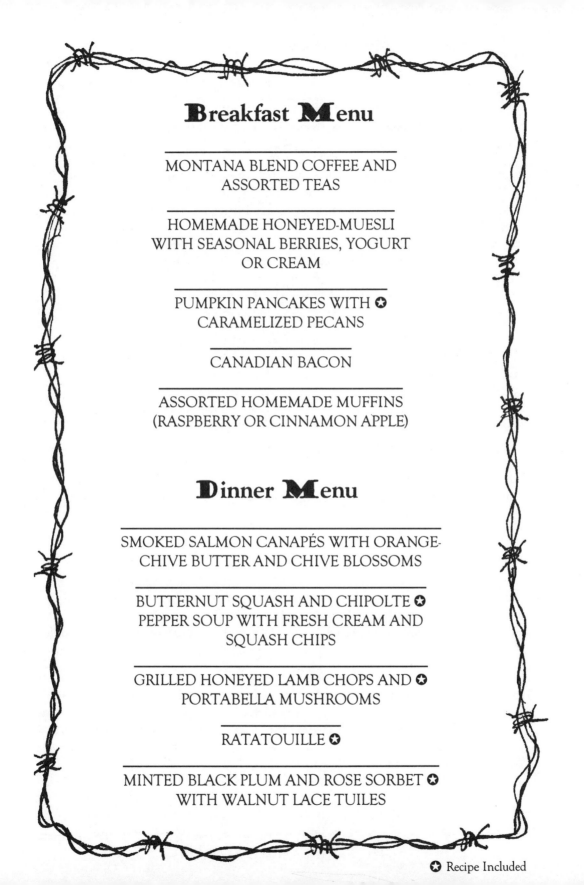
CARAMELIZED PECANS

CANADIAN BACON

ASSORTED HOMEMADE MUFFINS
(RASPBERRY OR CINNAMON APPLE)

Dinner Menu

SMOKED SALMON CANAPÉS WITH ORANGE-
CHIVE BUTTER AND CHIVE BLOSSOMS

BUTTERNUT SQUASH AND CHIPOLTE ✪
PEPPER SOUP WITH FRESH CREAM AND
SQUASH CHIPS

GRILLED HONEYED LAMB CHOPS AND ✪
PORTABELLA MUSHROOMS

RATATOUILLE ✪

MINTED BLACK PLUM AND ROSE SORBET ✪
WITH WALNUT LACE TUILES

✪ Recipe Included

Pumpkin Pancakes
with Caramelized Pecans

What a gorgeous golden-orange color! And so fluffy and thick, too. Our Bostonian friends Steve and Susan Lefkowitz were in town and dropped by just as I was popping these off the griddle. I dragged them into the kitchen and made them eat some. I didn't want to be the only witness to the unusual height and color of these pumpkins. The caramelized pecans are a perfect match to the earthy flavor and soft texture of the pancakes. I'm not a pumpkin pie eater, so these served at Thanksgiving would be perfect for me. I could still be all-American and have the pumpkin without the pie!

4-6 SERVINGS

1 cup whole wheat flour
2/3 cup all-purpose flour
2 teaspoons baking powder
1/4 teaspoon salt
2/3 cup pureed cooked pumpkin
    (or canned pumpkin)
2 eggs, separated
1 cup milk
2 teaspoons melted butter
1 tablespoon dark brown sugar

Caramelized Pecans:
1/2 cup pecans
1 tablespoon butter
Splash maple syrup

Sift flours, baking powder and salt. Beat pumpkin with yolks, milk, melted butter and brown sugar. Pour yolk mixture over flour mixture and stir. In a separate bowl, beat egg whites until they hold firm peaks. Fold a small amount of egg whites into batter, and then fold in the rest of the egg whites. (Doing this in 2 stages helps keep the batter light and airy, the key to fluffy cakes.) Ladle a scant 1/4 cup batter onto a hot buttered griddle or skillet. Cook until bubbles form on the surface and the edges barely start to dry. Flip cake and cook on the other side 1 to 2 minutes or until done.

To caramelize the pecans, heat the tablespoon of butter in a sauté pan until melted. Add the pecans and cook for 2 to 4 minutes, or until the nutty aroma is strong and the pecans darken just a little. Remove from heat and stir in a splash of maple syrup. I used a little more than a tablespoon, but add the syrup to your taste. Toss a few pecans on each pancake and serve with more maple syrup.

Butternut Squash and Chipolte Pepper Soup

An interesting flavor combination results through mixing the naturally sweet butternut squash and the smoky hot chipolte. Kathy, my tester, was so smitten with this soup she ate 2 bowls full. And she talked about it for days (and many recipes) afterward. The presentation is clever, too, consisting of dainty fresh cream swirls and crunchy squash chips.

8 SERVINGS

1/2 cup finely chopped celery
1 medium onion, finely chopped
1 teaspoon salt
1/4 cup butter (1/2 stick)
1 apple (preferably Braeburn), peeled, cored and chopped
1/2 teaspoon white pepper
1/2 teaspoon cracked black pepper
1 cup dry white wine
1 cup chicken stock*

2 medium butternut squash (2-1/2 to 3 pounds), peeled, cleaned and chopped
2 chipolte peppers (canned in adobo sauce) chopped
Milk for thinning if needed
Vegetable oil for frying squash chips
1/2 cup sour cream
1-3 tablespoons milk

In a large soup pot cook celery, onion and salt in butter over medium heat until tender. Add apple, white and black pepper, white wine and chicken stock. Turn heat to low and cover. Meanwhile, peel squash. With a vegetable peeler, shave 15 to 20 squash peelings (no skin) and set aside. Chop the remaining squash and add to soup pot. Cook until very tender, 45 to 60 minutes. While soup is simmering, heat about 1" oil in a saucepan over medium heat. When oil is very hot, add shaved squash peelings and cook until crisp and lightly browned. Remove from oil, drain on paper towels and season with salt and white pepper. Set aside.

When squash is tender, puree in batches in a blender or food processor. (Caution! Hot liquid in a blender can shoot the lid off. Hold the lid down tight.) Return soup to pan and stir in chopped chipoltes. Add a splash of milk to thin if necessary. Heat until hot enough to serve, but not boiling. To present soup, ladle 6 to 8 ounces in a decorative bowl. Mix sour cream and 1 to 3 tablespoons milk. Using a squeeze bottle or pastry bag with a small tip, zigzag lines across soup surface. Top with fried squash curl.

Concentrated chicken stock is available through More Than Gourmet. See the "Sources" section on page 236 for details. One (1.0-ounce) tin of Fond de Poulet Gold® makes 5 cups of chicken stock.

Grilled Honeyed Lamb Chops and Portabella Mushrooms

Red wine-soaked lamb sweetened with honey and grilled to perfection. Think it sounds ravishing? It is. This marinade was simply made for the unique taste of lamb. Alex and Rosalie, lamb-lovers at heart, sampled this delectable little treat and raved at the deep, complex flavors. The lamb needs to marinate for at least 12 hours, and is even better if you marinate it the night before. Chef Creagh's original recipe called for lamb loins. I couldn't get them on short notice, so I substituted lamb chops. I've called for chops in the recipe since that's what I tested.

4-6 SERVINGS

12 lamb chops (about 3 pounds)
4-6 large portabella mushroom caps,
    rinsed and dried

Marinade:
1/8 cup olive oil
1/4 cup red wine (preferably cabernet
    sauvignon)
1/8 cup balsamic vinegar
4 tablespoons honey
1 teaspoon dry mustard
Salt and pepper to taste

The evening before you plan to serve the lamb, mix all the marinade ingredients together. Place the lamb chops in a shallow pan and pour the marinade over the chops. Turn the chops over, cover and refrigerate. In the morning, turn the chops again and continue to refrigerate.

Chef Creah serves this dish with the Ratatouille (recipe follows), and if you do, too, then as you prepare the Ratatouille, let the marinating chops come to room temperature.

Preheat the grill to medium-high heat (375°-400°). Toss the portabella mushrooms in the marinade with the lamb. Grill the lamb and mushrooms. The lamb should take about 3 minutes per side for medium-rare (135°-140°). The mushrooms should be cooked until tender, about 5 minutes per side.

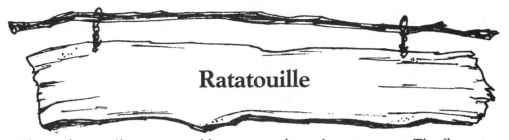

Ratatouille

This melange of hearty vegetables is covered in a fantastic sauce. The flavor is like a sweet tangy barbecue sauce only better. It very nicely complements the Grilled Honeyed Lamb dish. My friend Jill Rizzuto created this dish while I was working on the lamb. Her Italian heritage makes her think eggplant should be our national vegetable. I'm not an eggplant lover; so the only thing I would do different next time is cut the eggplant into smaller pieces and cook it a little longer than the other vegetables. I didn't change Chef Creagh's recipe to reflect my own personal taste, so you do what's best for you.

6-8 SERVINGS

1 medium eggplant cut into 2" cubes
1 red bell pepper, cut into 1-1/2" pieces
1 yellow bell pepper, cut into 1-1/2" pieces
1 zucchini, cut into 1/2" rounds
10 pattypan squash, cut in half (or 1 yellow squash, cut into 1/2" rounds)
1 red onion, cut in half lengthwise, then thickly sliced
1 + tablespoon olive oil
3 cloves garlic, finely chopped
2 bay leaves
3 sprigs basil (leaves only, discard stems)

Sauce:
1/2 teaspoon dried thyme
1/2 teaspoon white pepper
1 tablespoon fresh ground black pepper
2 tablespoons dark brown sugar
1/4 cup balsamic vinegar
1/2 cup red wine (preferably cabernet sauvignon)
1/2 cup tomato paste
Splash soy sauce
8 roma tomatoes, cut into 1/2" pieces
Garnish with basil sprigs, Parmesan cheese and kalamata olives

In a large sauté pan over medium high heat, add olive oil and cut vegetables. Season with salt and pepper. Cook until vegetables are nicely browned yet still firm. Transfer vegetable mixture to a large pot. Add garlic, fresh basil leaves, and bay leaves to vegetables and stir.

In a food processor, mix dried thyme, white and black peppers, brown sugar, balsamic vinegar, red wine, tomato paste and soy sauce. Process until just combined (a few pulses). Add to vegetable mixture and toss. Cook ratatouille on medium-low heat for 30 to 45 minutes, adding chopped roma tomatoes during the last 15 minutes of cooking.

Chef Creagh places the garnished ratatouille on a plate, and lays the lamb and mushroom on top for a beautiful presentation.

Minted Black Plum and Rose Sorbet with Walnut-Lace Tuiles

I'm not kidding. This is the most beautiful sorbet I've ever seen. The plums render a lovely rose color, and since you don't peel them, the sorbet is dotted with tiny specks of deep purple. Of course, looks alone do not make a worthy dessert. It has to taste good, too. And this sorbet does taste good. So good that my tester Kathy kept going back to the freezer "for just one more bite." My neighbor Rosalie also thought it was divine and ate her whole container in one sitting and wanted to know if there was more. The rosewater is subtle but you can taste it. It tastes just like a rose smells, delicate and floral. Kathy says to make the mint-infused sugar syrup the day before to give it time to develop and completely cool.

YIELD = 2 QUARTS

2-1/2 cups water
2-2/3 cups sugar
4 mint sprigs

10 ripe black plums, halved and pitted
1 teaspoon rosewater*

Heat water, sugar and mint sprigs in a small saucepan just until sugar melts. Remove from heat and chill (overnight preferably). Puree the plums and add rose water and chilled mint syrup (mint removed). Process in an ice cream maker or place in an airtight container and freeze until hard.

Walnut-Lace Tuiles

YIELD = 18 COOKIES

1/2 cup brown sugar
1/4 cup corn syrup
1/3 cup + 1 tablespoon butter

1/2 teaspoon salt
1/2 cup finely chopped walnuts
1/2 cup flour

Bring brown sugar, corn syrup, butter and salt to a boil. Add salt, flour and walnuts and stir well. Drop by teaspoonfuls (about 3" apart) onto a buttered cookie sheet and bake until golden brown, about 4 to 6 minutes. Watch carefully. Once they start to turn golden, it cooks quickly. The edges will brown first and then the color moves to the center. Cool only slightly and remove from pan. You can shape them while they are still very warm. Wrap them around the handle of a wooden spoon or drape them over small cups or glasses to create a cup shape. Experiment until you get the hang of it.

Rosewater is available through A.J.'s Fine Foods. See "Sources" section on page 236.

Triple Creek Ranch

5551 West Fork Stage Route
Darby, MT 59829
(406) 821-4600
Website: http://
    www.triplecreekranch.com
E-mail:
    triplecreekinfo@graphcom.com

Season: Year-round

Capacity: 42

Accommodations: 18 individual
cabins with fireplaces, stocked
wet bar, refrigerator, snacks,
telephone with dataport, TV/
VCR, bathrobes, hair dryers;
Relais & Chateaux member

Activities: Summer — horseback
riding; hiking; tennis; swimming;
fly-fishing; whitewater rafting;
Winter — skiing; snowshoeing;
guided snowmobile adventures;
historic or helicopter tours

Rates: $475-$995 per couple per
night, including all meals,
lodging and ranch activities

If you don't think southwestern Montana is a
mystical, magical, enchanting place, consider this
— a favorite activity of Triple Creek Ranch guests is
hooking up with a local realtor to find their own
slice of Bitterroot Valley paradise. Of course, most
of us just want to move into Triple Creek,
permanently. And no wonder. The property is a
member of the exclusive Relais & Chateaux and
has been described by the Robb Report as "the
best-kept secret in the Rocky Mountains." I can't
even begin to list all the glowing write-ups the
ranch has enjoyed over the years. I think one of
the most accurate, and my personal favorite, comes
from the Wall Street Journal, describing a stay at
Triple Creek as " . . . roughing it Robin Leach
style."

Designed as a romantic adult hideaway, nothing
has been left to chance. The quality-constructed
log and cedar cabins are appointed with glorious
western furnishings. An elegant central lodge with
an enormous picture window presents a
breathtaking bird's eye view of the West Fork of
the Bitterroot River. Romance is alive and well in
this secluded refuge where sipping champagne in
your hot tub, or snuggling up by the fireplace in
your cabin energizes your body and soul. Have
your dinner brought to your cabin, or dine by
candlelight in the intimate dining room. You'll feel
pampered and spoiled by the congenial staff and
the luxurious accommodations. If you can tear

yourself away from the alluring charm of your private cabin, the ranch offers a plethora of activities for you and yours to enjoy together. It doesn't matter if you visit in the summer or winter, the idyllic lure of the ranch is the same.

Summertime is filled with warm sunny days, perfect for hiking through wildflower-soaked meadows and thick pine-scented forests. Horseback riding is exhilarating, too, while exploring rugged trails, climbing to elevations that offer panoramic scenes you've only dreamed about. It's a great place to indulge in the art of fly-fishing, wading in crystal-clear streams where all you can hear is the rushing water, sparkling as it tumbles over time-worn rounded boulders. Float down the Bitterroot to access secret pools of black water teeming with plump trout. If golfing is more your style, there is a nearby gorgeous 18-hole course with astounding mountain views. The area offers a number of fine and Western art galleries and you can design your own custom cowboy hat in Wisdom, Montana. There is never a lack of things to occupy your time.

Winter brings a mystical, peaceful aura to the ranch. Watching the snowfall, listening to utter silence and traipsing through soft powder are favorite pastimes. If you want just a little more physical activity, there is cross-country and downhill skiing nearby. Or, cover miles on snowmobiles, viewing the winter wonderland. Return to the lodge for a soothing massage or a soak in the hot tub. Curl up in front of a crackling fire with a good book. There is a special quality to the ranch in the winter. It's solitude without being alone.

With the first class amenities and luxurious surroundings, you would think the food would be pretty good. It's not. It's far better than "good." It's phenomenal. Chef de Cuisine Martha McGinnis is a world-class chef-prodigy. Her experience as a chef aboard some of the finest cruise ships traveling the east coast, the Caribbean and Alaska along with extensive stints in famous Seattle restaurants have given Chef McGinnis a solid foundation in a variety of regional and international cuisine. Talented Sous Chef Nick Kormanick, trained at the Lederwolff Culinary Academy in Sacramento, California, has contributed to the esteemed culinary reputation at Triple Creek since the beginning of 1995. Together, the two chefs create a fabulous American regional fare that rivals any fine restaurant. Chef Carolyn Coyle manages the hearty breakfasts and varied and tasty lunches. In addition to her classical training, her extensive background in catering shines through with big, beautiful breakfast buffets and delightful packed lunches for those guests venturing out for the day.

Both the breakfast and dinner menus from Triple Creek are nothing short of outstanding. I think you will add these items to your repertoire and recreate the magic of Triple Creek over and over again.

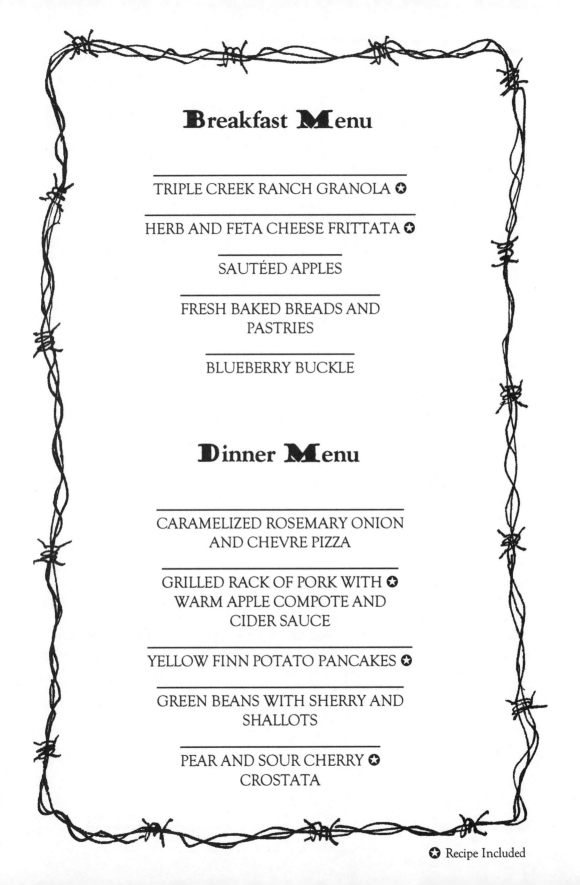

Breakfast Menu

TRIPLE CREEK RANCH GRANOLA ✪

HERB AND FETA CHEESE FRITTATA ✪

SAUTÉED APPLES

FRESH BAKED BREADS AND
PASTRIES

BLUEBERRY BUCKLE

Dinner Menu

CARAMELIZED ROSEMARY ONION
AND CHEVRE PIZZA

GRILLED RACK OF PORK WITH ✪
WARM APPLE COMPOTE AND
CIDER SAUCE

YELLOW FINN POTATO PANCAKES ✪

GREEN BEANS WITH SHERRY AND
SHALLOTS

PEAR AND SOUR CHERRY ✪
CROSTATA

✪ Recipe Included

Triple Creek Granola

This granola is so crunchy, nutty and blithely sweet. I kept a big bag of it beside my computer as I began writing this book and it didn't last through the first ranch. I could eat it all day long. I like the simple nut combination of almonds, hazelnuts and pepitas (the green inner seed within the pumpkin seed) with the high ratio of honey. I used dried cranberries and dried tart cherries, my favorite fruit combination for granola.

YIELD = 16-18 CUPS

1 (18-ounce) container old-fashioned oats
1-1/2 cups sliced almonds
1-1/2 cups pepitas*
1-1/2 cups hazelnuts, roasted, skinned and coarsely chopped
1-1/2 cups sweetened coconut
1-1/4 teaspoon cinnamon
1/2 cup + 1 tablespoon light brown sugar
1-1/2 cups honey or maple syrup
3/4 cup vegetable oil
2 cups chopped dried fruit

Preheat oven to 350°. Toss oats, nuts, coconut, cinnamon and brown sugar until well-mixed. Heat the honey and oil just to warm. Pour over dry ingredients and mix until all ingredients are coated with honey mixture. Spread mixture onto lightly greased sheet pans with edges. Bake, stirring frequently to brown evenly (about 15-25 minutes). Keep a close eye on the granola as toward the end of browning, it turns very quickly. Remove from oven, add dried fruit and stir occasionally to break up clumps, and cool. Store in airtight containers or sealable plastic bags. Keeps for weeks well-stored. Freeze for storage longer than 8 weeks or in very humid weather.

*Pepitas are available through A.J.'s Fine Foods. See "Sources" section on page 236 for details.

Herb and Feta Cheese Frittata

Delicious! This is light and fluffy and bursting with fresh herb flavors. The chopped tomato adds a bright splash of color and a pleasing acidic taste.

1 SERVING

1 teaspoon butter, melted
1 tablespoon chopped green onion
1 tablespoon finely chopped yellow onion
1 tablespoon finely chopped fresh herbs*
1/2 cup chopped fresh spinach
3 eggs, beaten and seasoned with salt and pepper
1/4 cup crumbled Feta cheese
1 tablespoon seeded and finely chopped tomato

Brush an 8" non-stick skillet with melted butter. Heat the pan and add the green onion, yellow onion, fresh herbs and spinach. Cook until tender, about 3 or 4 minutes. Pour beaten eggs over vegetables, filling pan until evenly distributed and scrambling lightly with a rubber spatula for the first minute or two. When bottom is set, sprinkle with Feta and place pan under broiler or in oven until puffed and golden. Slide the frittata onto a cutting board and cut into wedges. Place on a warm plate and garnish with chopped tomato. Serve immediately.

*The ranch recommends an herb mixture of rosemary, parsley, thyme, chives and sweet basil. We used parsley, tarragon and basil because that is what we had that was fresh. It was a great combination, too.

Grilled Rack of Pork with Apple Compote

Chef McGinnis has really outdone herself with this recipe. I am in awe of her ability to concoct such a fabulous combination of flavors and taste sensations within one dish. The most tragic event during my recipe testing is preparing this dish late in the evening with no testers and no tasters. However, I'm sure the people in the next county heard my exclamations of oohs and ahs, and wows. I did save the leftovers for Kathy, my tester, to try the next day. Even as a leftover this dish receives our highest marks. Ask your butcher to cut the pork rack into individual chops for you. (Packaged, pre-cut chops are just not as plump and juicy as fresh ones, and besides, your butcher will love the opportunity to cut specially for you. If not, change butchers.)

5-6 SERVINGS

Apple Compote:
1 Rome apple, peeled, cored and chopped
2 small Golden Delicious apples, peeled,
    cored and chopped
3/4 cup dried tart cherries

3/4 cup chopped dried apricots
1 cup brown sugar
1 tablespoon finely chopped fresh sage
1/2 cup cider vinegar
1/2 cup unfiltered apple cider (or juice)

Combine all ingredients in a medium saucepan and simmer slowly, stirring occasionally, until almost all the liquid has been absorbed and the compote has a dark, syrupy consistency. (It took me about 45 minutes with a 2-1/2 quart saucepan.) May be prepared 1 day in advance.

Continued next page

Apple Cider Sauce

1/4 cup finely chopped shallots
1 tablespoon finely chopped garlic
1/4 cup apple liqueur (Applejack)
2 cups unfiltered apple cider (or juice)

6-8 fresh sage leaves, chopped
2 cups Demi-Glace Gold®*
3 cups Glace de Poulet Gold®**
Salt and pepper

In a small, shallow sauté pan, combine shallots, garlic and apple liqueur. Set aflame with long-handled match or lighter. After flame dies, add cider and chopped sage and cook 10-15 minutes or until roughly 1/2 cup of liquid is left. Add demi and chicken glaze and simmer until it thickens and coats the back of a spoon without dripping, about 20 to 30 minutes. Strain, discarding solids. May prepare 1 day in advance up to this point. Before serving, heat and season with salt and pepper.

Grilled Pork

Rack of pork, 5-8 ribs, cut into individual chops
Salt and pepper

Heat grill to medium-high heat (375°-400°). Mark chops by placing on hot grill and cooking 3 minutes, then turning chop 25° to the right, same side. Cook another 3 minutes and reduce heat to 350°. Grill to medium temperature (140°). Let rest 5 minutes, loosely covered with foil. To serve, place 1/3 cup warm compote on a warm plate. Lay grilled chop on top, at an angle, with bone pointing up. Ladle 1/4 cup sauce over lower 1/3 of chop.

*Demi-Glace Gold® is available through More Than Gourmet. See "Sources" section on page 236 for details. One (1.5 ounce) tin of Demi-Glace Gold® makes 1 cup of demi-glace.

**Glace de Poulet Gold® is available through More Than Gourmet. See "Sources" section on page 236 for details. One (1.5-ounce) tin of Glace de Poulet Gold® makes just over 1 cup.

Yellow Finn Potato Pancakes

Can there be such a thing as a bad potato dish? No. But there are good potato dishes and there are fantastic potato dishes. This one is fantastic! So tasty and crunchy. Although the name of this recipe calls for Yellow Finn potatoes, we were equally successful with plain russet potatoes. Using Yellow Finns or Yukon Golds adds a little color and perhaps a smidgen more of flavor. But those varieties of potatoes are not always available and I refuse to have to wait until they are to enjoy this crispy treat.

4-6 SERVINGS

1 generous pound scrubbed potatoes (Yellow Finn, Yukon Gold, or Russet)
1/2 medium yellow onion, peeled
2 tablespoons finely chopped fresh chives
1/4 teaspoon fresh ground pepper
1/2 teaspoon salt
1-1/2 tablespoons all-purpose flour
1 egg beaten with 2 tablespoons heavy cream
1/4 cup melted butter (1/2 stick) + 1/4 cup vegetable oil mixed together

Coarsely grate unpeeled potatoes and onion into a colander or strainer. Rinse thoroughly with cold water until all the potato starch is rinsed away and the water runs clear. Place the potato mixture, a handful at a time, in a clean dishtowel. Gather up the sides of the towel to make a pouch and squeeze the potatoes until no more liquid comes out. Put dried potato mixture in a bowl.

Preheat oven to 350°. Toss potato mixture with chives, pepper, salt and flour. Add egg and cream mixture and toss to combine. Heat a sauté pan or cast iron skillet over medium high heat. Add butter mixture to a depth of 1/8" (make more butter mixture if necessary for your pan, 1/2 melted butter and 1/2 vegetable oil). Make sure pan is hot then drop potato mixture into pan in 1/3 cupfuls. Squish with a spatula to flatten it out and cook until golden brown, turn and cook other side until golden brown. Place on a baking sheet and finish in a 350° oven for 5 to 7 minutes.

Pear and Sour Cherry Crostata

I think this is a gorgeous pie. So rustic with its folded edges and flat shape.
It's super easy to put together. Sometimes simplicity can be sophisticated.
The taste is delicious, just the right balance between the sweet pears and tart
cherries. Use your own favorite pie crust dough or one of the recipes in this
book. We used the pie dough from Lone Mountain on page 178.

8-10 SERVINGS

1 recipe for 9" single pie crust
1 cup dried sour cherries
3 tablespoons dark rum
4 ripe pears, peeled, cored and sliced 1/2" thick
3/4 cup sugar
2 tablespoons butter
2 tablespoons heavy cream
2 tablespoons sugar

Steep the cherries in a sauté pan with the rum on low heat until plumped, about 15
minutes. Melt butter in a large non-stick sauté pan and add pears, cherries and sugar.
Toss to combine and cook for about 5 minutes until the pears are just barely tender.
Remove to a bowl and chill.

Preheat oven to 350°. Roll out piecrust on a lightly floured surface to a circle of 11"-12".
Lift onto a greased sheet pan. Mound chilled pear and cherry filling onto dough, leaving
a 2" border all the way around. Fold the border over the filling, pleating as you go. (A
good portion of the filling will be uncovered.) Brush the top of the dough with the cream
and sprinkle heavily with sugar. Bake 30 to 40 minutes or until pastry is lightly browned
and cooked through. Let rest 5 to 10 minutes before cutting. Serve with a dollop of
whipped cream or vanilla ice cream.

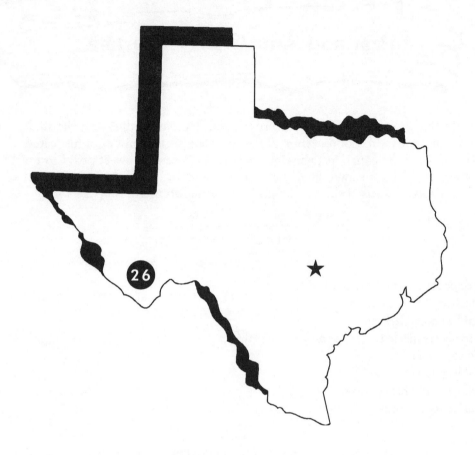

TEXAS

Cibolo Creek Ranch

P.O. Box 44
Shafter, TX 79850
(915) 229-3737
Website: http://
     www.cibolocreekranch.com

Season: Year-round

Capacity: 32 overnight guests

Accommodations: The ranch has 3 separate historical forts with a variety of guestrooms from ultra-luxurious to a quaint guest cottage. The forts have been painstakingly restored to reflect the original structure down to the last nail.

Activities: Horseback riding; hiking; mountain climbing; four-wheeled tours, historical sites; skeet shooting; white-water rafting; cooking classes

Rates: $250-$590 per night, double occupancy, including meals and non-alcoholic beverages

CIBOLO
CREEK RANCH

Enveloping 25,000 acres, Cibolo has become a favorite stomping ground for celebrities and sophisticated travelers alike. Mick Jaggar and crew are repeat guests at the ranch. One look at the place and you can see why it's earning a reputation as an exclusive hideaway. Situated deep in West Texas, it's not the easiest place to get to, but then that's the point for most of its guests. The ranch has its own airstrip and the nearest airport, albeit small, is 60 miles away in Alpine, Texas. For major airline carriers, you fly into El Paso or Midland. Once you get to the ranch, you're not ever going to want to leave. Serenity, Texas hospitality and incredible food from a rising culinary star provide guests with too many reasons to stay.

West Texas terrain is high desert, dotted with sage, chaparral and ancient cottonwood trees fed by natural springs. The landscape is so stark and enchanting that several high fashion magazines have produced photo shoots on the ranch, including <u>Vogue</u> and England's <u>Frank</u> magazine. There are so many things to do both at the ranch and nearby. Try exploring Indian caves and discovering Native American artifacts, or visiting a ghost town, or four-wheeling through the backcountry of the ranch for incredible crimson sunset vistas. The Big Bend National Park and McDonald's Observatory are also nearby as is

Mexico (20 miles south). Many guests have great intentions to pack their week with exciting activities only to melt into the relaxed atmosphere and leisurely pace at the ranch. They end up just absorbing the sun by the alluring pool, or taking a nap in shaded hammock, dreaming about their next trip to Cibolo and all the activities they'll do next time. Or maybe they are just dreaming about Chef Ahier's next spectacular feast!

Culinary Institute of America graduate Chef Ahier is a force to be reckoned with in her field. Her ingenuity and raw talent are helping her make a name for herself and are bringing culinary star status to Cibolo. She calls her style "Cross-cultural cuisine." I call it phenomenal. She revels in mixing indigenous ingredients with ethnic influences, all within the realm of classical French cooking techniques. She uses local organic products when at all possible and if she can't get the best, she won't serve it. Menus are not planned too far in advance so that Chef Ahier can take advantage of local harvests. "I just cook. That's it. I'm into fresh, organic food. I'm more concerned about the taste of my dish than I am about the presentation. It's just food after all," says Chef Ahier. Her modesty is refreshing and her food is glorious. Enjoy the menu selections Chef Ahier has shared with us and discover the taste of Cross-cultural cuisine.

Breakfast Menu

BUTTERMILK WHOLE WHEAT
BISCUITS

BROILED TEXAS RUBY RED ✪
GRAPEFRUIT WITH CITRUS HONEY
AND TOASTED COCONUT

FLORENTINE BREAKFAST PIZZA ✪

Dinner Menu

DILL AND TEQUILA SHRIMP SALAD ✪
(CIBOLO STYLE) WITH IOWA BLUE
CHEESE DRESSING

MAPLE CHILI GLAZED QUAIL ✪

CANADIAN WILD RICE WITH ✪
DRIED CRANBERRIES, CHERRIES
AND TOASTED PECANS

TRIPLE CHOCOLATE CAKE

✪ Recipe Included

Broiled Texas Ruby Red Grapefruit with Citrus Honey and Toasted Coconut

Sweet and tart, this Texas treat is best when grapefruits are in season, October through February. Even non-grapefruit fans will devour this warm, spritzy dish. You can use any pink grapefruit, but Ruby Reds are Lisa's favorite. See the "Sources" section on page 236 for how to order authentic Texas Ruby Reds from Frank Lewis' Alamo Fruit company.

8 SERVINGS

4 Ruby Red grapefruit (or pink grapefruit)
1 cup citrus honey (or your favorite honey)
1 cup shredded coconut

Slice grapefruits in half crosswise. Run a knife around the outside of the fruit and through the sections for easier eating and to create pockets for the honey to fall. Drizzle grapefruits with 2 tablespoons of honey each and top with 2 tablespoons of coconut. Broil 5 minutes, or until warm and coconut is golden brown, being careful not to burn coconut.

B. Hillis

Chef Ahier's Healthy Pizza Dough

YIELD = 5-8" PIES (5 OUNCES EACH),
OR 3-12" PIES (8 OUNCES EACH)

1 cup warm (110°) water
1 tablespoon honey
1 tablespoon dry yeast
2-1/2 cups all-purpose flour
1/2 cup whole wheat flour
1/2 cup cornmeal
1/2 teaspoon salt
1/4 cup olive oil

To make yeast mixture, fill a 2 or more cup measuring cup with 1 cup of warm water. Stir in honey until dissolved. Gently add yeast, stirring until all yeast is incorporated. The yeast mixture should be light tan with no lumps. If it is not, throw it out and start over. This is very crucial to a light and crispy dough. Cover and let sit for about 10 minutes or until a puffy foam forms on the top of the mixture.

Combine the all-purpose and whole wheat flours, cornmeal, salt and olive oil in a large mixing bowl. Stir until a sticky ball forms. (You may want to add either a little more all-purpose flour or water, depending upon the consistency before you knead the dough.)

Turn the dough out onto a floured surface and knead until dough becomes smooth and elastic. It is impossible to over-knead this dough, so knead until you are tired of kneading and then turn the dough into an oiled bowl, turning once to coat all sides of the dough. (You can do the kneading in a mixer with a dough hook if you prefer). Cover with plastic wrap and then a towel, and set in a warm spot and let rise for 1 to 2 hours or until dough doubles in size. Punch dough down to remove all air bubbles. At this point you are ready to use your dough. It can be stored in a sealed plastic bag in the refrigerator for up to 24 hours or in the freezer for up to 3 months.

Florentine Breakfast Pizza

This pizza is colorful as well as delicious. I think the picante sauce is crucial for a full-flavored pie. You could make this pie anytime of day, not just breakfast. I think it would be a great early evening weekend meal, paired with a Fume Blanc or Beaujolais.

YIELD = FILLING FOR 3-8" PIES

1 pound fresh spinach, washed (about 8 cups loosely packed)
4 eggs
1 cup ricotta cheese
1 cup picante sauce (optional)
2 cups shredded mozzarella cheese

Boil spinach in unsalted water until cooked thoroughly, about 15 minutes. Drain well and squeeze out all excess moisture. (If there is any moisture left in the spinach, your pizza will be soggy, so take some time with this).

Preheat oven to 500°. Place spinach, eggs and ricotta in a food processor or blender and blend for 2 minutes. Roll out dough into an 8" circle and place on a lightly oiled perforated pizza pan. (Chef Ahier prefers Airbake brand pans.) Spread picante sauce over dough, if using. Pour spinach mixture on top of pizza and sprinkle with mozzarella. Bake for 15 to 20 minutes, until crust is medium brown.

Dill and Tequila Splashed Cibolo Shrimp Salad

with Iowa Blue Cheese Dressing

If you bought this book for only one killer recipe, this might be the one. This is a knock-out salad with flavors bursting off the plate. Cibolo Shrimp is an upscale, updated version of the 80's Buffalo Wings craze, and a fine example of the depth of talent of Chef Ahier.

8 SERVINGS

40 pieces large shrimp, completely shelled, de-veined and chilled
3 cups buttermilk
1 cup all-purpose flour
3 cups all-purpose flour
2 teaspoons salt
1 teaspoon ground black pepper
1 teaspoon ground red pepper (cayenne)
Oil for frying, about 2 cups
2 heads romaine or butter lettuce, washed, dried and torn into bite-size pieces

Sauce:
1/2 cup butter (1 stick)
1/4 cup finely chopped roasted garlic
1/2 cup tequila
1 cup good quality red pepper sauce (but not Tabasco)
1/4 cup Tabasco
1/4 cup chopped fresh dill (do not use dried)
Iowa Blue Cheese Dressing (recipe follows)

While shrimp and lettuce are chilling in refrigerator, make sauce by melting butter in skillet over medium heat. Add garlic, tequila, red pepper sauce, Tabasco sauce and dill. Whisk over medium heat for 2 minutes. Remove from heat and set aside.

Mix 3 cups buttermilk and 1 cup flour in a large bowl. Stir in shrimp. Heat 2 cups or more oil in a large skillet over medium heat (350°). In a roasting pan or other shallow pan, mix 3 cups all-purpose flour, salt, black and red pepper. Using a slotted spoon, transfer a few pieces of shrimp at a time to the roasting pan and dust the shrimp with the flour/pepper mixture. Shake off excess flour and fry shrimp until golden brown, about 5 minutes, and drain on paper towels.

Toss lettuce in a large bowl with enough dressing to barely coat lettuce. Divide dressed lettuce among 8 chilled salad plates.

Toss drained shrimp in sauce and arrange 5 shrimp on top of each salad plate and serve immediately. Use remaining blue cheese dressing as dipping sauce for shrimp.

Iowa Blue Cheese Dressing

It's called Iowa dressing because Chef Ahier uses Maytag Blue Cheese from Iowa. If you can't find it locally, you can order it through A.J.'s Fine Foods. See the "Sources" section on page 236. I personally think it's the best blue cheese, but if you have another favorite, by all means use it instead.

YIELD = 2 CUPS

1/2 cup sour cream
1/2 cup mayonnaise
1/2 cup blue cheese
1 teaspoon Tabasco sauce
1/2 teaspoon Worcestershire sauce
1/4 teaspoon salt
1 tablespoon lemon juice
2-4 tablespoons milk (to thin)

Place all ingredients except milk in a food processor or blender and blend until combined. Thin with milk to desired consistency. Chill.

B. Hillis

Maple Chili Glazed Quail

Here's another recipe that's putting Chef Ahier on the national culinary map. Creative, innovative and uniquely Cibolo, this quail is incredibly simple to make and intensely flavorful. The mild heat from the chili powder teamed with the pure sweetness of maple syrup will have you asking why didn't you think of combining the two contrasting flavors. Use the freshest chili powder and the best grade of maple syrup you can find.

8 SERVINGS

16 semi-boneless quail
1/4 cup vegetable oil (canola preferred)
2 cups maple syrup
1/4 cup chili powder
Salt and pepper to taste

Preheat grill to medium-high heat (375°-400°). Rinse and dry quail. Lightly coat quail with oil and salt and pepper. Combine syrup and chili powder in a small saucepan and bring to a boil. Remove from heat and let cool to thicken. Preheat oven to 400°.

Grill quail for 2 minutes per side. Place grilled quail on a baking sheet and brush maple glaze all over quail. Bake 2 minutes and re-brush with glaze. Cook another 2 to 3 minutes, or until thoroughly cooked.

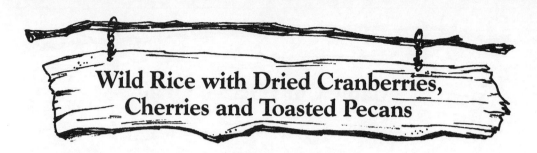

Wild Rice with Dried Cranberries, Cherries and Toasted Pecans

I like the contrasting flavors and textures. The pecans are crunchy and the fruit is chewy, with the rice texture somewhere in between. The herbaceous flavor of the wild rice is complemented by the sweet tartness of the cranberries and cherries. It's a perfectly balanced dish. Chef Ahier also likes to serve this as a cool refreshing summer treat by sprinkling the rice mixture over baby field greens with simple citrus vinaigrette.

8 SERVINGS

1 cup wild rice, uncooked
4-5 cups vegetable stock or water
1/4 teaspoon salt
3 scallions, sliced thinly
1/4 cup dried cranberries
1/4 cup dried cherries
1/4 cup toasted pecans*
1/4 cup cranberry and orange juice mixed
1 tablespoon finely chopped orange or lemon peel
Salt and pepper to taste

Rinse rice in cold water until water runs clear. Combine rice and stock or water and salt in a heavy-bottomed saucepan. Bring to a boil and reduce heat to a simmer. Cover and cook about 45 to 65 minutes or until rice is tender, adding more stock or water if necessary. Rinse and drain rice.

Mix rice with dried fruit, pecans, juices and orange or lemon peel. Season to taste with salt and pepper. Reheat on stovetop or in microwave if necessary.

*To toast nuts, see page 11.

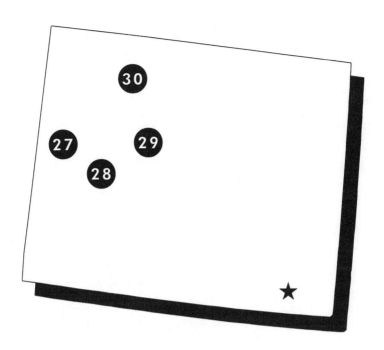

WYOMING

Crescent H Ranch

P.O. Box 347
Wilson, WY 83014
(307) 733-3674
(888) TETONS1 (838-6671)
Website: http://
www.crescenth.com

Season: mid-June through mid-September

Capacity: 25

Accommodations: 10 rustic but elegant log cabins, 70+ year-old main lodge

Activities: Premier fly-fishing; horseback riding; gourmet dining; guided hiking trips; weekly BBQ; rodeo; Grand Teton and Yellowstone National Parks

Rates: $350 per night/double occupancy; fly-fishing package - $2,700 per person per week, includes 7 nights, 5 guided days, all meals and fly-fishing instruction if desired; no credit cards

CRESCENT H RANCH

Just a short drive from Jackson, Wyoming, is the famous Crescent H Ranch. Built in 1927, the lodge embodies the spirit of western hospitality as much today as it did more than 70 years ago. Noted for spectacular fly-fishing the Crescent H also has a history of providing outstanding horseback riding and unparalleled hiking opportunities. Nearby are the Grand Teton and Yellowstone National Parks. Within the town of Jackson, there are over 30 art galleries and museums, and Jackson is now considered a mecca for western art.

The Crescent H Ranch has the distinct honor of being the first Orvis-Endorsed Lodge. Orvis, the largest fly-fishing products manufacturer and the oldest cataloger in North America, established the first endorsement program ever in 1983. The purpose of the program is to seek out those facilities located in exceptional fishing areas and whose accommodations, amenities and service are first-rate. The endorsed lodges are continually reviewed by the Orvis staff and must meet the rigorous standards to maintain this honor. Of the many hundreds of guest ranches worldwide, only 39 are endorsed by Orvis. It's no wonder the Crescent H was the first endorsed lodge. The fishing is extraordinary, with over 7 miles of private blue ribbon water on the ranch reserved for guests only. There are also several spring creeks on the property, providing some of the most challenging fly-fishing you're likely to encounter anywhere. With crystal clear, near motionless water, and a dry fly barely bigger than a gnat, mastering spring creek fishing is enormously rewarding. Fishing off the ranch property is world-famous. Most serious fly-fishers have heard of and dreamed of fishing the renowned Snake River or the legendary Yellowstone River. Take advantage of the

ranch's proximity to this stellar fishing water and the seasoned, professional guides provided by the ranch.

Miles of trails on the ranch appeal to all skill levels of horseback riders, and there are morning and afternoon rides everyday with the ranch's experienced wranglers. During the summer, Jackson is home to one of the world's most famous rodeos, with events every Wednesday and Saturday night. The Jackson Hole area is also famous for its hiking possibilities. Hundreds of hiking trails in and around the area vary from light to moderate to serious terrain hiking. There is plenty of hiking on the ranch itself, or just a short distance away in the Grand Teton National Park. You can also choose to do nothing but rest and relax back at the ranch. The best place to "hang" is the large wooden porch off the main lodge in a comfortable rocker, soaking up the scenery and breathing the fresh clean air. It's a good place to swap fish stories at the end of the day or plan the next day's adventure.

Anticipating another exquisite meal from Executive Chef Dan Ripley and his talented staff is a popular ranch activity as well. Dining at the Crescent H is always a highlight and it's no wonder with the amount of time and preparation the kitchen applies to each and every meal. One of the things that impresses me the most about the culinary scene (besides the fact that the food is stellar), is the experience of the culinary team itself. Chef Ripley has been with the ranch for 10 years and the Sous Chef and Pastry Chef have logged 4 years each. With a team that's been together that long, the execution of creative cuisine becomes flawless. Each travel in different directions in the off-season and return to share new ideas, new techniques and create new exciting menu items to add to the old favorites. Chef Ripley's enthusiasm for the ranch as a whole and not just his culinary realm struck me as rare but heartening. Is he the Chef or the ranch's goodwill ambassador? He's both. He is extremely knowledgeable about the entire operation and clearly loves being a part of this organization.

During your week's stay, you will be exposed to a wide selection of wild game, fresh fish, local beef and free-range birds. How does Roasted Duck Breast with Sun-dried Cranberry Sauce sound? Or how about Smoked Chicken Pasta with Tomatoes, Garlic, Basil and Pine Nuts? The Grilled Filet of Beef with Cabernet Plum Sauce is one the guests' favorites. Chef Ripley utilizes his herb garden and forages fresh produce from local farmers' markets. The breads and pastries created by Chef Sean Beck receive as many kudos from guests as the main course. Chef Beck churns out fresh croissants, brioche, breakfast pastries as well as rustic artisan breads daily. Somehow, he still has time to create desserts that are as beautiful as they are delicious. You'll be tempted with a Banana Rum Torte, or a White and Dark Chocolate Mousse Cake. Chef Beck always provides a choice of two desserts. (I'd get both if I were you. You're on vacation, after all!)

Breakfast at the Crescent H is anything you want. You can have eggs, bacon, sausage, and toast or pancakes or waffles, or omelets or anything else you can think of. Once a week guests ride out to a lush meadow beside one of the spring creeks and enjoy a "Cowboy Breakfast" cooked outdoors. Chef Ripley describes the dinner menu he shared with us as a French-influenced California style, with fresh, wholesome ingredients.

Dinner Menu

BUTTERNUT SQUASH AND
GINGER SOUP

Or

CARAMELIZED ONION TART ✪
WITH GORGONZOLA SAUCE

CRISPY SEA BASS WITH A LEMON ✪
DILL BEURRE BLANC

SWEET POTATO CAKES

SAUTÉED SPINACH

CHOCOLATE CHOCOLATE CHIP ✪
ICE CREAM

✪ Recipe Included

Caramelized Onion Tarts with Gorgonzola Sauce

After I made this I called Chef Ripley and told him I thought he was a genius. I'm seeing this sauce on all kinds of dishes. It's so creamy, but not heavy, and has just the right twang from the cheese. I'm thinking beef tenderloin with this sauce or steamed asparagus or herb-stuffed chicken, or, well, you get the picture. It's fabulous on these puff pastry rounds, too.

6 SERVINGS

Onion filling:
6 Vidalia* or other sweet yellow onions
1/2 cup butter (1 stick)
2 tablespoons fresh thyme
Salt and pepper

Thinly slice the onions. In a large sauté pan, melt the butter and cook the onions until caramelized (deep golden brown color), about 25 to 35 minutes. Stir in fresh thyme and season with salt and pepper.

Tart shells:
1 sheet puff pastry or 6 puff pastry circles
1/2 cup water
1 egg

I bought the package of puff pastry circles. There were 6 to a box and they were 3" in diameter, which is what you need for this recipe. If you buy the sheet, cut out 6 circles with a 3" cookie cutter. Mix water and egg together to make a wash. Brush circles with the egg wash and place on a lightly greased sheet pan. Bake until golden brown, about 20-25 minutes. Cool and then hollow a 2" circle out of the center. Set aside.

Gorgonzola sauce:
1-1/2 cups heavy cream
1 cup crumbled Gorgonzola cheese (blue cheese)
1 tablespoon chopped fresh thyme

Bring cream to a boil. Reduce heat and add cheese. Add thyme and season with salt and pepper. Cook another 2 to 4 minutes until the consistency is thick enough to coat the back of a spoon.

Putting it all together: Place a puff pastry shell on a plate. Place about 1/4 cup of caramelized onions on the inside and spread a few around the outside of the shell. Drizzle 2 to 3 tablespoons of sauce all over plate making sure a good portion is on the tart shell. Garnish with a sprig of fresh thyme.

I don't think any other sweet onion compares to a Vidalia. If you've not tried one, you've got to get your hands on one. I listed a source on page 237. You can't get them any fresher than direct from the source where they are grown.

Crispy Sea Bass
with Lemon Dill Beurre Blanc

Doesn't "Beurre Blanc" sound so much better than "melted butter"?
Actually, it's more than just melted butter. It's a classic French sauce, based
on a white wine reduction and whole butter. And there are as many ways to
make this sauce as there are cooks in kitchens. Chef Ripley's choice of lemon
and dill, two classic fish accouterments, with the ultimate butter sauce is
perfect for this expensive aquatic creature. (Apparently, it's extremely labor-
intensive to get a fresh piece of fish in Arizona. Maybe you live closer to the
source than I do.) Sea Bass is really delicious, and this simple preparation is
the best way to have it.

6 SERVINGS

6 (5-ounce) boneless sea bass filets
    (preferably Chilean sea bass)
Flour for dusting
1/4 cup olive oil
Salt and pepper to taste
Sauce:
1 shallot, finely chopped
1 teaspoon olive oil

1-1/2 cups white wine
1 bay leaf
6 whole black peppercorns
3/4 pound butter (3 sticks) cut into pieces
    and at room temperature
Grated peel of 1 lemon
1 tablespoon chopped fresh dill

Season sea bass with salt and pepper. Lightly dust both sides of the sea bass. (It's best to
do this by sprinkling the flour over the fish with your fingers, as opposed to laying the
fish in the flour. You want only a bare amount of flour.) In a large skillet, heat olive oil
over medium-high heat until very hot. Carefully add filets to the hot oil and cook until
they turn golden brown and crispy. Flip over and crisp other side. Remove from heat and
finish in a 400° oven (about 4 to 5 minutes).

To make the sauce, heat olive oil in a medium sauté pan and add shallots. Cook 1
minute and add wine, bay leaf and peppercorns. Cook 15 to 20 minutes, or until most of
the liquid is evaporated (you only need about 1 to 2 tablespoons of liquid left in the pan).
Remove from heat and whisk in butter at room temperature until it is completely
incorporated. Strain sauce and stir in grated lemon peel and dill. Taste and season with
salt if necessary. Keep covered and warm until ready to use. Careful not to heat too
much, or the sauce will "break" (fat will gather on the surface).

Place a hot sea bass filet on a warm plate and spoon 1 to 2 tablespoons of sauce on top
and around fish. Garnish with a sprig of fresh dill and/or a lemon wedge.

Chocolate Chocolate Chip Ice Cream

Can it be? Something so overtly rich and chocolaty that my neighbor Alex said "please, no more?" I can't believe it. It's definitely rich. Too rich? You'll have to decide. It is, without a doubt, the creamiest, chocolatiest ice cream I've ever tasted.

YIELD = 1-1/2 PINTS

2 cups half and half
1/2 vanilla bean
4 ounces semi-sweet chocolate, chopped
8 egg yolks (yes, eight!)
2/3 cup sugar
1/2 cup semi-sweet chocolate morsels

In a medium saucepan, heat the half and half, vanilla bean and the 4 ounces of semi-sweet chocolate over medium heat until the chocolate is melted. Increase heat to medium high and bring chocolate mixture just to a boil. Remove from heat. In a small bowl, beat eggs and sugar until thick and lemon colored. Add a small amount of the hot chocolate mixture to the egg mixture and stir. Add another small amount of the chocolate mixture to the egg mixture and stir. Now that the egg mixture is warm, add all the egg mixture to the chocolate mixture. (This process is called "tempering." By bringing the temperature of the egg mixture up, it won't "cook" when you add it to the hot chocolate mixture.) Return the combined mixture to the stove and cook over medium heat until the custard coats the back of a spoon and does not drip (this won't take long). Strain mixture and chill over an ice bath, stirring frequently. When cool to the touch, cover and refrigerate the custard for at least 1 hour. Process in an ice cream maker. Store in airtight containers in the freezer for up to 2 weeks.

Flying A Ranch

Route 1 Box 7
Pinedale, WY 82941
(307) 367-2385 (Summer Season
    June-October)
(605) 332-0946 or
(800) 678-6543 (November-May)
Website: http://
    www.guestranches.com/
    flyinga

Season: Mid-June through
October

Capacity: 12-14 (Adults Only)

Accommodations: Six completely
renovated rustic cabins, named
after early valley settlers

Activities: Guided horseback
riding; fly-fishing; hiking;
mountain biking; hot tub

Rates: June 14-August 30, Sunday
to Sunday- $1,175 per person
double occupancy, $1,675 single
occupancy

August 30-October 4, Sunday to
Sunday- $1,100 per person
double occupancy, $1,600 single
occupancy

Nestled between the Gros Ventres and Wind
River mountain ranges, just 50 miles southeast of
Jackson Hole, Wyoming, lies the picturesque
Flying A guest ranch. The ranch is located 7 miles
off the highway and the drive up to the ranch
affords guests the opportunity to begin unwinding,
passing pastoral meadows and soaking in the
enfolding majestic mountain ranges. The ranch
sits at 8,300 feet above sea level, though you
wouldn't think you were that high after seeing the
surrounding mountain peaks.

Historic cabins brought up to date with all the
modern conveniences help make you feel you've
stepped back in time without giving up current
"necessities." The cabins all have original hand-
carved native pine furniture on top of rich new oak
floors. The cabins' porches are cozy and perfect for
relaxing at the beginning or end of the day, while
the wildlife laggardly grazes in the lush meadow
surrounding the ranch.

The adults-only format of the Flying A Ranch
offers guests the opportunity to relax in peace and
quiet, without the hustle and bustle of everyday
city life. With less than 15 guests each week,
Debbie and her staff of 10 genuinely pamper and
spoil the guests, making them wish for just one
more week. Many guests book for the following
year before they even leave the ranch, and the

Flying A boasts a guest repeat rate well above 60%. A guest from Kansas wrote, "Thanks again to each of you for making our experience at the Flying A such a great one. Everything was so well done — the delicious food, service, the care of the cabins, the trail rides and hikes and the friendliness of all the staff was excellent. On a scale of 1-100, you all rate a 200!"

Guests dictate how active they want to be while at the ranch. Choosing from an extensive array of activities, most guests take advantage of some of the 28 horseback trails and Debbie says not to miss Whiskey Peak, where the elevation is close to 10,000 feet, or Waterdog Lakes and Jack Creek Falls trails. Whichever trail you take, you can be sure that your wrangler will guide you through cool groves of aspen and pine for some remarkable scenery that is tough to rival. If riding is not on your agenda, there is fly-fishing (with lessons included) on 2 large, stream-fed ponds on the property. Or, let the ranch guides take you to a local river, including the world-famous Snake River. Guides also offer hiking and mountain biking excursions. Nearby are Yellowstone and Grand Teton National Parks, the Green River and Green River Lakes, Jackson Hole (western shopping "Mecca") and the Mountain Man Museum in Pinedale. The best thing about planning your week of activities at the Flying A is you decide how much or how little you want to do.

After a full day of activity, nothing is more rewarding than returning to the ranch for an evening of Epicurean delights. Debbie blends Western flair with continental cuisine and the resulting fusion is nothing short of exquisite. The Flying A is becoming known for its casual elegance, not only in the accommodations but also in the culinary area as well. Every single meal is presented with a different place setting and served on fine china. With an experienced Chef and Baker on board, the Flying A offers guests some of the finest food you will find anywhere.

The presentation of the food is equally as important as the flavor and both get lots of attention from the kitchen staff. Many of the fresh vegetables you'll savor during your week come from the greenhouse on the ranch. Fresh smoked trout tops the guests' most treasured appetizer and all steaks are cut at the ranch to ensure the quality and consistency Debbie demands. Special dietary needs are not a problem and the staff is more than willing to accommodate your particular needs. Debbie has shared two of her favorite menus and some of the accompanying recipes.

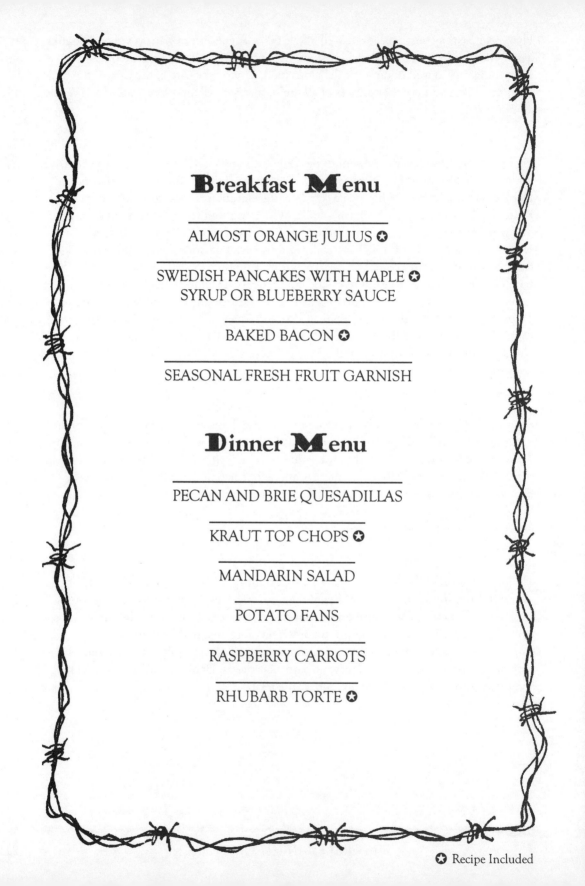

Breakfast Menu

ALMOST ORANGE JULIUS ✪

SWEDISH PANCAKES WITH MAPLE ✪
SYRUP OR BLUEBERRY SAUCE

BAKED BACON ✪

SEASONAL FRESH FRUIT GARNISH

Dinner Menu

PECAN AND BRIE QUESADILLAS

KRAUT TOP CHOPS ✪

MANDARIN SALAD

POTATO FANS

RASPBERRY CARROTS

RHUBARB TORTE ✪

✪ Recipe Included

Almost Orange Julius

Yummy! This drink tastes just like that orange and vanilla ice cream bar I ate when I was young; so creamy and delicious. You'll have this recipe memorized in no time at all.

4-6 SERVINGS

1 (6-ounce) can frozen orange juice
   concentrate
1 cup milk
1/4 cup sugar (optional)

1/2 teaspoon vanilla extract
1 cup water
10 ice cubes

Mix all ingredients in blender until ice cubes are thoroughly crushed. Serve immediately.

Baked Bacon

This is a great way to prepare bacon and the brown sugar adds an unusual crunchy sweetness. The wafting aroma will bring everyone into the kitchen. I used peppered bacon for an extra spice kick.

6 SERVINGS

12 slices thick-sliced bacon
1/2 cup brown sugar

Dip bacon slices in brown sugar. Place on rack of a broiler pan or on a rack in a shallow pan. Bacon should not lie in drippings. Bake at 375 ° for 20 to 25 minutes, or until bacon is crisp.

Swedish Pancakes

These pancakes are gorgeous! So thin and delicate looking, they turn a deep golden brown when cooked with beautiful markings from the cooking process. The buttermilk provides a tangy flavor that's not easy to forget.

6-8 SERVINGS

2 eggs
1 teaspoon salt
1 tablespoon sugar
2 cups flour

1 tablespoon baking powder
1 teaspoon baking soda
3 cups buttermilk
6 tablespoons butter, melted

Beat eggs in large bowl. In a small bowl, mix salt, sugar, flour, baking powder and baking soda. Add this flour mixture and the buttermilk to the beaten eggs. Beat well and stir in the melted butter. Preheat griddle or skillet over medium heat. Pour 1/4 to 1/3 cupful of batter onto griddle. Cook until top surface is bubbly and edges lightly browned. Turn and cook until other side is lightly browned. Serve with maple syrup, blueberry sauce or your favorite syrup.

Blueberry Sauce

YIELD = 3 1/2 CUPS

4 cups fresh or frozen blueberries
1/2 cup sugar
2 tablespoons cornstarch
1 cup water

In medium saucepan, combine sugar and cornstarch. Gradually stir in water. Add blueberries; bring to a boil over medium heat. Boil for 2 minutes, stirring constantly. Remove from heat; cover and keep warm until served.

Kraut Top Chops

I have a little confession to make. I didn't know what "Kitchen Bouquet" was until I tested this recipe. If you don't know either, it's a browning and seasoning sauce and I found it next to the barbecue sauces in the grocery store. It makes this dish quite tasty and I can see lots of other uses for it as well. My husband is not a sauerkraut fan and he really enjoyed these chops, leaving only the bone on the plate. The breadcrumbs make a nice crunchy topping.

4-6 SERVINGS

6-8 thick pork chops. (If you do not want the bone, use cuts from a pork loin or roast)
Kitchen Bouquet
8-10 ounces sour cream
1 (15-ounce) can sauerkraut
1/2 cup breadcrumbs
2 cups beef stock
Parmesan cheese
Dash of paprika

Brush pork chops with Kitchen Bouquet. Place in shallow baking dish. Cover pork chops with sauerkraut. Blend sour cream with 2 cups beef stock and pour over sauerkraut-topped chops. Sprinkle with breadcrumbs and Parmesan cheese. Sprinkle with paprika. Bake for 1 hour at 350°.

Rhubarb Torte

My good friend Jill made this dessert and I love the way she presented it. After it was baked, she scooped servings into little ramekins and it looked just like a cobbler instead of a pie. The taste is more cobbler-like, too, tart and biscuity. Yummy! A big dollop of vanilla ice cream on this warm pie is even better.

8-10 SERVINGS

For the crust, combine:
1 cup flour
Pinch of salt
1/2 cup butter (1 stick), cut into small chunks
5 tablespoons powdered sugar

Preheat oven to 375°. Pat crust into a 9" X 9" pan. Bake for 12 minutes.

For the filling, combine:
2 eggs, beaten
1-1/2 cups sugar
1/4 cup flour
3/4 teaspoon baking powder
3 cups diced rhubarb (can use frozen)

Place filling in pre-baked pie shell and bake another 35 to 40 minutes more. Serve with fresh whipped cream or vanilla ice cream. Delicious!

T Cross Ranch

Box 638
Dubois, WY 82513
(307) 455-2206
Website: http://www.tcross.com

Season: June through October

Capacity: 24

Accommodations: Lovely log cabins nestled in the pines; western décor including lodgepole furnishings, Indian rugs and artifacts; woodstoves/fireplaces

Activities: horseback riding; hiking; fly-fishing; campfire singalongs; pack trips; horseshoes; tubing down Horse Creek; hot tubbing

Rates: $1,050 weekly per person; includes all meals, lodging and ranch activities

T CROSS RANCH

If you have ever wanted to travel back in time, to see how the old west really was, step through the gates of T Cross Ranch. Although you will be quite comfortable with all the modern amenities, you will feel as if you have been transported back to another era. A fugitive from the Johnson County cattle wars homesteaded the ranch in the late 1800's and by 1920 T Cross was operating as a guest ranch. The ranch is owned and managed by Ken and Garey Neal, who work hard to maintain the authenticity of the original ranch. The small number of guests each week in tandem with the secluded location of the ranch provides a welcomed intimacy. Tucked away in a lush meadowed valley next to the Shoshone National Forest, the pristine 160-acre ranch is dotted with grassy meadows, tall pines and a meandering creek. Only a half-hour north of the cowboy town of Dubois, Wyoming and only two hours away from Jackson, Wyoming, T Cross is the perfect place to escape to feel isolated from the rest of the world.

Part of the reason T Cross is so popular with returning guests is the easy pace and low key atmosphere at the ranch. While there are plenty of activities to keep you busy if you want, it seems as if there is plenty of time for jumping on a horse, or grabbing a fly-rod and heading to Horse Creek or Wiggins Fork. No one seems to be rushed to pack in a week's worth of activities. The ambiance is very relaxing and peaceful. When you are ready for horseback riding, the ranch's wrangler will pair you up with a horse suited to your skill and you can ride mornings, afternoons or all day. Trails will take you into deep timber, shimmering aspen groves and up to spectacular vistas. The fly-fishing is superb, both on the ranch's Horse Creek and nearby fabled Wiggins Fork.

If you want to venture out from the ranch, there are many local attractions to tempt you. The Greater Yellowstone Area and the Grand Teton National Park are within driving distance as is Jackson, Wyoming with shops and galleries and museums. The Washakie Wilderness is right next door. Dubois, Wyoming, half an hour away, is still a small town with an abundance of western charm and full of cowboy ways. Most guests tend to want to stay within the intimate embrace of the ranch and snuggle up in front of the fire in their cabin or in the main lodge. Wafting smells from the kitchen lure guests to the cozy dining room.

Food at T Cross is wholesome, plentiful and always delicious. Breakfast begins with hot coffee or tea and an assortment of fresh juices and fruits. Guests are offered their choice of eggs any style, pancakes or waffles, all cooked to order. Each morning there is also a specialty item such as an egg dish or fresh baked coffeecake or pastry. Guests gather in the main lodge around 6:00 p.m. for cocktails and hors d'oeuvres. Dinner follows shortly after and guests are treated to hearty comfort food with gourmet touches. The plates always look elegant, a nice balance of color and texture. Most importantly, the food on the plate is delicious and filling. There is always plenty so ask for seconds if you want. T Cross shares one of its breakfast specialty items and a lovely dinner menu with the best carrots I've ever eaten.

B.Hillis

THE GREAT RANCH COOKBOOK

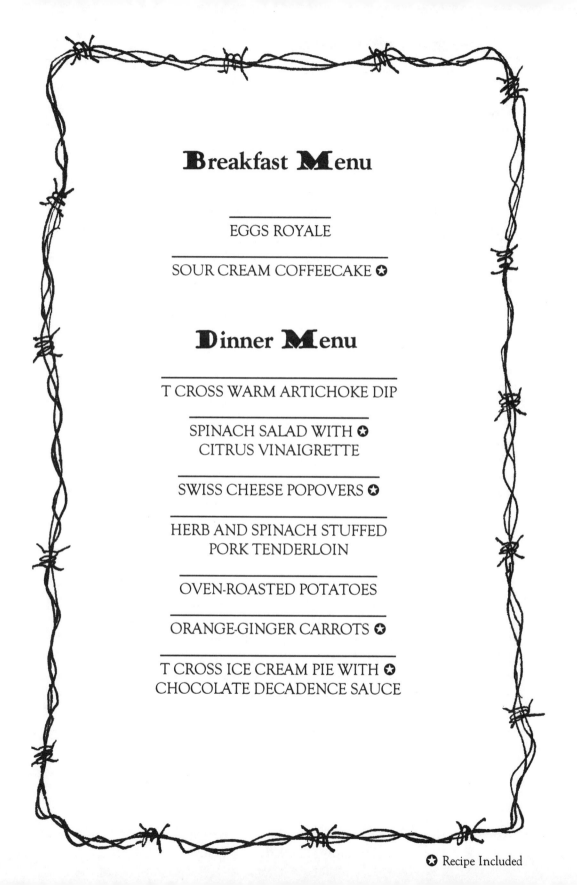

Breakfast Menu

EGGS ROYALE

SOUR CREAM COFFEECAKE ✪

Dinner Menu

T CROSS WARM ARTICHOKE DIP

SPINACH SALAD WITH ✪
CITRUS VINAIGRETTE

SWISS CHEESE POPOVERS ✪

HERB AND SPINACH STUFFED
PORK TENDERLOIN

OVEN-ROASTED POTATOES

ORANGE-GINGER CARROTS ✪

T CROSS ICE CREAM PIE WITH ✪
CHOCOLATE DECADENCE SAUCE

✪ Recipe Included

Sour Cream Coffeecake

Marilyn Robertson was so gracious during the recipe testing, eating what she could, and distributing the rest to friends and family. She tasted nearly everything in this book. Quite an accomplished cook herself, not once did she ask me for a recipe until she tasted this coffeecake. She liked the moist, dense texture and the pronounced cinnamon and nut flavors. The filling for this cake makes a pretty swirl through the cooked cake.

8-10 SERVINGS

1 cup butter (2 sticks), softened
2 cups sugar
2 eggs
1 teaspoon vanilla
1 cup sour cream
2 cups flour
1 teaspoon baking powder
1/4 teaspoon salt

4 teaspoons brown sugar
1 cup chopped pecans
1 teaspoon cinnamon

Preheat oven to 350°. Beat butter and sugar until smooth. Add eggs and beat again. Add vanilla and sour cream and mix well. In a another bowl, mix flour, baking powder and salt. Stir flour mixture into sour cream mixture until blended. Grease and flour a 10" fluted tube pan and pour in 1/2 batter. Mix brown sugar, nuts and cinnamon together. Sprinkle 2/3 of nut mixture over batter. Top with remaining batter. Top with remaining 1/3 nut mixture. Lightly run a knife through the nut mixture, incorporating it into the batter slightly. Bake for 50 minutes or until a toothpick inserted in the center comes out clean. Remove and cool in pan for 10 to 15 minutes. Turn out onto cake rack and cool. May serve warm or cold.

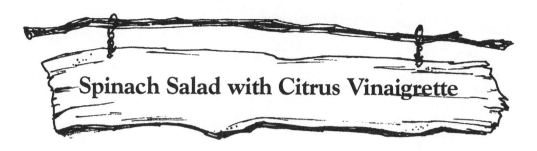

Spinach Salad with Citrus Vinaigrette

The dressing on this salad is to die for. There is no oil in the dressing. It's made from the bacon drippings, which is why it probably is one of the most flavorful I've ever tasted.

6-8 SERVINGS

2 bunches spinach
10 slices bacon, cut in slivers
1/3 cup bacon drippings
1/4 cup orange juice
1/4 cup cider vinegar
1/4 cup currant jelly
3/4 cup sliced almonds, toasted*
1 (11-ounce) can mandarin oranges

Rinse, trim and dry spinach. Tear into bite size pieces and place in a large bowl. Fry bacon until crisp and drain on paper towels. Reserve 1/3 cup drippings in a small saucepan. Add orange juice, vinegar and jelly to bacon drippings. Just before serving, bring dressing to a boil and pour over spinach. Set aside a handful of bacon and toasted almonds for garnish, adding the remainder to the salad and toss. Garnish with mandarin oranges and reserved bacon and almonds.

*To toast almonds, see page 11.

Swiss Cheese Popovers

These popovers aren't baked in the traditional muffin-like popover pans, but they look like rustic popovers and taste divine. A crunchy crust and an airy, moist inside characterize these cheesy treats. They are easy to prepare and look and taste unique and different.

YIELD = 16-18 POPOVERS

1 cup water
6 tablespoons butter
1 teaspoon salt
1/8 teaspoon pepper
1 cup flour
4 eggs
1 cup shredded Swiss cheese

Preheat oven to 400°. In a medium saucepan, bring the water, butter, salt and pepper to a boil. Add the flour all at once and stir vigorously until the mixture forms a ball that pulls away from the sides of the pan and leaves a starchy film on the bottom, 2 to 5 minutes. Remove from heat. One at a time, beat in the eggs, fully incorporating before adding the next. Stir in Swiss cheese. Using a 2" scoop or 2 heaping tablespoons, drop batter onto lightly greased baking sheet. Bake 40 minutes, or until puffy and golden brown. Turn oven off and leave door open with popovers inside for 15 minutes. Remove and serve warm.

B. Phylis

Orange-Ginger Carrots

I'll admit I'm not a carrot fan. Never have been and probably never will be. However, I have received some great carrot recipes for this book. I have to rank this one at the top. The carrots are not so much covered in a "glaze" as they are sauced with a "gel." An intriguing and luscious gel at that. As Patty Carter, one of the owners, was handing over the recipes for this book, she said, "You're going to love the gingered carrots." Boy, she wasn't kidding. I think I could even ask for seconds.

6 SERVINGS

9 medium carrots, peeled
1 tablespoon sugar
1 teaspoon cornstarch
1/4 teaspoon salt
1/4 teaspoon ground ginger
1/4 cup orange juice
1 tablespoon butter
Grand Marnier (Optional)
1 tablespoon chopped parsley

Cut carrots on the bias into 1/4" thick slices. Steam over boiling water until done to your desired texture (we did 5 to 6 minutes for crisp-tender). Meanwhile, in a small saucepan, cook the sugar, cornstarch, salt, ginger and orange juice until thick and bubbly. Reduce heat. Stir in butter and simmer one more minute to get the gel-like consistency. Taste and add more ginger or Grand Marnier if desired.

When carrots are done, drain. Pour sauce over carrots and toss. Garnish with fresh parsley.

T Cross Ice Cream Pie
with Chocolate Decadence Sauce

Awesome and easy to make! You can buy a chocolate crust to make it even easier. The sauce makes this more than just an ice cream pie. It takes it to another level. The chocolate sauce makes great sundaes, too. I prefer the orange whisper of Grand Marnier over the plain vanilla version.

8-10 SERVINGS

Crust:
25 chocolate wafers, crushed,
    about 1-1/2 cups
6 tablespoons butter, melted

Filling:
1 pint coffee ice cream
1 pint chocolate ice cream
1/2 cup chopped pecans
1/2 cup chopped chocolate-toffee candy
    bars

Pull ice cream from freezer to soften. Preheat oven to 350°. To make the crust, crush the cookies to fine crumbs and mix with butter. Press into a 10"pie plate and bake 10 minutes. Remove and completely cool. While the crust is cooling, make the chocolate sauce (recipe follows).

When the ice cream is softened, mix ice cream, pecans and chopped candy in a large mixing bowl. Pour into cooled crust and freeze until hard. To serve, cut pie into 8 to 10 slices, place on a chilled dessert plate and drizzle pie and plate with chocolate sauce.

Chocolate Decadence Sauce

YIELD = 2 CUPS

4 ounces semi-sweet chocolate, chopped
1 cup heavy cream
1 cup sugar
3 tablespoons butter
1 tablespoon vanilla or Grand Marnier

Over simmering water in a double boiler, melt chocolate and stir in cream. Mix to combine, stirring constantly. Add sugar and butter and stir until sugar is dissolved. Remove from heat and stir in vanilla or Grand Marnier. Allow to cool at room temperature (refrigerating will cause sugar to crystallize and you'll have a grainy sauce).

UXU Ranch

1710 Yellowstone Highway
Wapiti, WY 82450
(307) 587-2143
(800) 373-9027
Website: http://
www.uxuranch.com
E-mail: uxuranch@aol.com

Season: June through September

Capacity: 36

Accommodations: 11 cozy, comfortable cabins with private porches, some overlooking river and some with gas fireplaces

Activities: Horseback riding with instruction; guided fly-fishing; mountain biking, hiking; windsurfing; guided trips to Yellowstone National Park; rodeos; children's program

Rates: $1,325 for 1st adult (14 years and up), $1,050 for each additional adult weekly per person; children ages 3-5, $525; ages 6-13, $900

R A N C H

I don't know if I can do justice, describing the feeling one experiences, standing in the meadow in front of the UXU's main lodge. It's part awe, part communion. Towering pine trees, swaying gently and scenting the air, occupy the foreground while gigantic, rugged mountain faces stand majestically in the background. It's a contrasting sensation, one of feeling like an invited guest of Mother Nature while at the same time feeling a sense of ownership in a land time forgot. UXU is a special place, and a treasure chest of experiences. Located in central Wyoming, only 17 miles to the east of Yellowstone National Park, the ranch is just off the road that Teddy Roosevelt called the most scenic 50 miles in America. I think he would still feel the same way traveling this road again today.

If you can take your eyes off the scenery long enough, there is a multitude of activities in which to partake. Horseback riding will allow you to continue feeding your eyes with luscious scenery. Take the mid-week all day ride that climbs over 2,000 feet. While the ride is challenging, the views are incredible and unmatched. The resident fly-fishing guide is more than happy to refresh your skills or teach you new ones, and show you some secret fishing spots. The famous waters of Yellowstone National Park, like its namesake, the Yellowstone River, the Firehole and the Madison, recognized as superior western waters for beautiful rainbow and native cutthroat trout, are within easy reach from the ranch. Surrounded by the Shoshone National Forest, the number of hiking and biking trails are practically unlimited. For something different, try windsurfing in the nearby Buffalo Bill Reservoir that Outside Magazine called "one of the best windsurfing lakes in the country." Trap and skeet shooting and golf are also nearby. Take a trip into town and visit the Buffalo Bill Historic Center, courtesy of the ranch.

Even with all of these spectacular activities, one that remains a guest favorite is meandering through the ranch property, following old June Creek or strolling through the lush green meadows spotting various wildlife or unusual wildflowers. Before you know it, afternoon has turned to dusk and dusk to dark. It's a chance to collect your thoughts, rejuvenate your soul and harmonize with the wonder of nature. One guest summed it up best with "we loved the way the week matured into a gathering of friends, beneath more stars than we ever realized exist in the universe."

Guests first travel to UXU because of the abundance of activities, the pristine setting and warm hospitality of owner Hamilton "Ham" Bryan and his remarkable staff. They often return with an added reason — the fabulous food. Staying true to a Wyoming flavor, the Executive Chef capitalizes on the abundance of fresh local produce and meat to provide fresh regional cuisine with classic presentation. Breakfasts are hearty and there is always a healthful option. Emphasis is on fresh fruits and homemade fresh baked goods. Dinner is a time to relax and leisurely indulge in multi-course affairs. All sauces, salad dressings and relishes are homemade as are the dinner breads and fabulous desserts. You have a choice of nightly entrees and special diets are always accommodated. Rumor has it that Ham is quite the cook as well, and has been known to whip up a dish or two on occasion. He keeps his secret recipes close to the vest, but generously shared some ranch favorites with us.

B. Hillis

THE GREAT RANCH COOKBOOK

Breakfast Menu

FRESH JUICE AND FRUITS

MULTI-GRAIN PANCAKES WITH ✪
TOASTED COCONUT AND
BANANA COMPOTE

MAPLE CURED SAUSAGE

ASSORTMENT OF FRESH BREADS
AND PASTRIES

Dinner Menu

SPINACH SALAD WITH CHUTNEY
MUSTARD DRESSING

GRILLED ALBACORE TUNA WITH ✪
RED PEPPER SAUCE

GARLIC MASHED NEW POTATOES

GARLIC FOCACCIA ✪

STEAMED CHOCOLATE PUDDING ✪
WITH DRIED CRANBERRIES

✪ Recipe Included

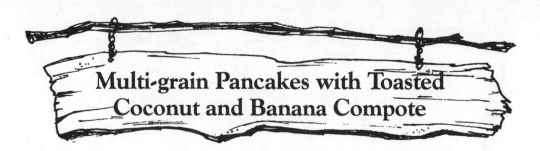

Multi-grain Pancakes with Toasted Coconut and Banana Compote

Multi-grain is an understatement. There are so many different grain sensations going on in this pancake that it's fun to try to single out the taste of each one. (It can't be done, but try anyway, and yes, we culinary-types are easily amused!) The compote is nice, especially if you like toasted coconut. We still added maple syrup to the pancakes at the table and it didn't detract from the compote. In fact, I think it enhanced and elevated the compote flavors.

YIELD = 15 (4") PANCAKES (ABOUT 4-5 SERVINGS)

3/4 cup rolled oats
1/3 cup wheat flour
1/3 cup cornmeal
1/3 cup buckwheat or rye flour
2 teaspoons baking powder
3/4 teaspoon baking soda
1/4 teaspoon salt
2-1/2 cups buttermilk
1/8 cup honey
1 tablespoon butter, melted
1 egg, beaten

Compote:
2 tablespoons brown sugar
1/4 cup toasted coconut
4 ripe bananas
1/4 teaspoon lemon juice
3 tablespoons butter

Mix the first 7 ingredients (oats through salt) together in a large bowl. In another large bowl, mix the buttermilk, honey, melted butter and egg. Pour the buttermilk mixture over the oats mixture and stir to combine. Batter will be thin. Heat a lightly greased skillet or griddle over medium heat and drop batter by 1/4 cupful. Cook until bubbles pop on surface and edges just start to dry, about 3 or 4 minutes. Flip over and cook 2 or 3 minutes, or until golden brown.

To make the compote, peel and cut the bananas into chunks. Melt butter and sugar in a skillet over medium heat. Add the bananas and cook for a couple of minutes. Add lemon juice and stir. Spoon over pancakes and sprinkle with toasted coconut. You can toast the coconut in a skillet over medium heat for 5 to 7 minutes, until it turns golden brown or in a 350° oven on a sheet pan for 5 to 7 minutes.

Garlic Focaccia

The garlic flavor dominates this crunchy-on-the-outside, soft-on-the-inside bread. Two keys to making this the best it can be are to let the dough really rise (way above the pan edges) during the second rising and to make the indentations with your fingers as called for in the recipe. After you make it this way, try adding sautéed onions or mushrooms or sliced tomatoes or chopped herbs on top before the second rising for a different bread.

YIELD = 16 PIECES

2 cups warm water (110°)
1/4 cup yeast
5 cups all-purpose flour
3/4 cup olive oil + 2 tablespoons to brush top
2 tablespoons finely chopped garlic, divided
1 tablespoon salt

Mix warm water and yeast. Let stand 10 minutes. Add 3 cups flour, 3/4 cup olive oil, 1 tablespoon garlic and salt. Mix with wooden spoon until smooth. Add remaining flour 1/2 cup at a time. Turn dough out onto a lightly floured surface and knead for about 5 minutes. Place in oiled bowl and cover. Let rise in a warm place until doubled in size, about 40 to 60 minutes. Turn dough out onto a work surface and punch out air. Roll or press into a cookie sheet (11-1/2" X 17"). Mix the remaining 2 tablespoons of olive oil with the remaining tablespoon of garlic and brush on top of bread. Cover and let rise. Preheat oven to 400°. Make a few indentations in the bread with your fingers. Bake about 20 to 25 minutes or until golden brown.

Grilled Albacore Tuna with Red Pepper Sauce

Stop! Even if you don't like cilantro, don't just skip over this recipe. The cilantro is very subtle and extremely complementary to the fish and the marinade when grilled. It might even make you change your mind about this pungent herb that flavors so many Mexican and Asian dishes. Albacore is harvested off the coast of Mexico and our West Coast. You can use any species you'd like, but Albacore is reasonably priced and I think not as strong-tasting as other varieties.

6 SERVINGS

1/2 cup olive oil
1/4 tablespoon lemon juice
4 tablespoons finely chopped cilantro
2 cloves garlic, finely chopped

1/2 teaspoon salt
1/4 teaspoon white pepper
6 (6 to 8-ounce) tuna filets, 1-1/2" thick
Red Pepper Sauce (recipe follows)

Whisk the oil, lemon juice, cilantro, garlic, salt and white pepper until thick and smooth. Place the filets in a single layer in a shallow pan. Pour oil/lemon mixture over fish and then turn fish over to coat the other side. Cover with plastic wrap and marinate for 1 hour in the refrigerator. Preheat grill to medium-high heat (375°-400°). Grill tuna 10 minutes for each 1" thickness, turning half way through cooking. (If your steak is 1-1/2" thick, grill for 15 minutes, turning over after 7 minutes.) Place on a warm plate and drizzle with Red Pepper Sauce (recipe follows).

Red Pepper Sauce

This sauce contains raw egg yolks, and should be handled with care. After making the sauce, refrigerate it and keep it in the refrigerator until ready to use. After saucing the fish, serve immediately. You can keep the sauce 1 day in the refrigerator.

1 red bell, seeded and cut into 8 pieces
1 jalapeno, seeded and chopped
3 cloves garlic, chopped
2 egg yolks

1 tablespoon fresh lemon juice
1/2 teaspoon salt
1 cup olive oil
2-3 tablespoons bread crumbs, optional

In a blender, combine the red pepper, jalapeno, garlic, yolks, lemon juice and salt. Blend until very smooth, about 2 minutes. With the motor running, very slowly pour in the olive oil in a steady stream. If you'd like it thicker, add the bread crumbs and blend again.

Steamed Chocolate Pudding with Dried Cranberries

Normally I don't like to adulterate my chocolate with anything. Period. But I'm willing to make an exception here. The small amount of cranberries is delicious. They kind of settle to the bottom and add just a tiny bit of texture to the otherwise smooth-as-silk pudding. The chocolate flavor is divine, rich and deep, but the consistency is light and squishes in your mouth (that's a good thing). It's best served warm, and will look like a flat cake. It can be eaten cold, and it resembles a flourless cake (torte).

8-10 SERVINGS

9 ounces semi-sweet chocolate
2 sticks + 2 tablespoons butter
5 ounces strong coffee*
2 tablespoons all-purpose flour
1/4 teaspoon salt
2-1/4 cups sugar, divided
5 eggs, separated
1/4 cup dried cranberries

Preheat oven to 350°. Melt butter and chocolate with coffee over low heat. Add 2 cups sugar, salt and flour. Beat egg yolks with 2 tablespoons sugar until thick and add to chocolate mixture. Whip egg whites with remaining 2 tablespoons sugar to stiff peaks. Add cranberries to half of egg whites and fold into chocolate mixture. Now fold in the remaining half of egg whites. Butter a 9" springform pan. Pour in batter and wrap completely and tightly with foil, also buttering the piece that covers the top of the pudding. Set a large roasting pan in the oven. Set the springform pan in the roasting pan and fill with hot water, about half way up the side of the springform pan. Bake for 1-1/2 hours.

Remove from oven. Remove springform pan from roasting pan, remove foil and cool 10 minutes. Cut into wedges and serve.

*My favorite coffee comes from a small specialty roaster in Austin, Texas, called Little City Roasting Co. They have a wide selection of coffees, and some special beans from East Africa. See "Sources" section on page 236 for details.

Sources

A.J.'S FINE FOODS

A.J.'s is a specialty grocery market chain in the Phoenix area. They carry most anything you can imagine. I relied on the Pima Road location and its manager Ron Blankenship for the hard-to-find items like pepitas, almond oil, achiote paste, etc. Butcher manager Bob Stachel located all the wild game I used in testing the recipes for this book. Call Ron, or any of the other A.J's stores and I guarantee you'll be pleased with the quality and service.

23251 N. Pima Road
Scottsdale, AZ 85255
(602) 563-5070

7141 E. Lincoln Drive
Scottsdale, AZ 85253
(602) 998-0052

5017 N. Central Avenue
Phoenix, AZ 85012
(602) 230-7015

13225 N. 7th Street
Phoenix, AZ 85022
(602) 863-3700

10105 E. Via Linda #110
Scottsdale, AZ 85258
(602) 391-9863

FRANK LEWIS' ALAMO FRUIT

Royal Ruby Red Grapefruit, oranges, tangelos, mangos, persimmons, watermelon and nuts.

100N. Tower Rd.
Alamo, TX 78516
(800) 477-4773

LITTLE CITY ROASTING CO.

Fresh roasted coffee blends, including custom blends. Top Shelf Blend is most popular and my favorite. Owner Donna Tayor-DiFrank also carries exotic East African beans you're not likely to find elsewhere.

3403 Guadalupe
Austin, TX 78705
(512) 467-2326
(512) 459-9670 (fax)

MORE THAN GOURMET

Classic French sauce bases — Glace de Poulet Gold®; Demi-Glace Gold®; Fond de Poulet Gold®; Jus de Poulet Lié Gold®; Veggie-Glace Gold®. All sauces are made with the finest and freshest ingredients and contain no chemicals, MSG or preservatives.

115 West Bartges Street
Akron, OH 44311
(800) 860-9385
(330) 762-4832 (fax)
E-mail: demi-glace@worldnet.att.net

PECOS VALLEY SPICE COMPANY

Quite possibly the most complete, freshest source for chile cooks interested in pure, quality products for the very best in southwestern cooking. Try the comprehensive quality line of authentic masas, herbs, spices and even seeds for growing chiles, tomatillos and the like. Check out the cookbooks and videos by famous southwestern cookbook author Jane Butel.

P.O. Box 964
Albuquerque, NM 87103
1-800-473-TACO (8226)
(505) 247-1719 (fax)
E-mail: pvsco@aol.com
Website: http://www.janebutel.com

VIDALIA ONION STORE

Sweet Vidalia onions. These are large, sweet onions. These onions are mild enough to eat like an apple.

Box 1719
Vidalia, GA 30475
(800) 226-3359

About the Author

Gwen Ashley Walters graduated, with honors, from the Scottsdale Culinary Institute. She sharpened her culinary skills at a number of restaurants and at the world-class Boulders Resort in Carefree, Arizona. She has managed an award-winning Orvis-Endorsed fly-fishing lodge in Southwestern Montana and is a frequent guest ranch visitor.

For 13 years before deciding to follow her heart into the culinary field, Ms. Walters was associated with a major marketing services company where she rose to the rank of Vice President. She graduated from North Texas State University with a Bachelor of Business Administration degree and a Master of Education degree.

She lives in Scottsdale with her husband and two dogs, Georgia and Taylor.

The Cast

I could not have completed this dream, this book, without the help of some very special stars. Thanking them is not enough; I am indebted to them, for their contributions and their love and friendship. And now you are the beneficiary of the expertise and strength of these remarkable individuals as this book is the result of their collective effort, and my good fortune in knowing them. I want to share with you just a bit of the joy these people brought me during this extraordinary episode of my life. I wish you could have been here.

THE WISE ONES:

Jeff Walters, husband and best friend

This book was his idea. I wanted to write a cookbook and he channeled my desire in this direction and supported me in a myriad of ways. He tasted recipes on the few occasions he was home and traveled the rest of the time, staying out of the way. He read chapters for content and technical errors and generously shared his knowledge of authorship. (His book is Measuring Brand Communication, ROI, published by the Association of National Advertisers, 1997). He is a good person and a great man and I'm lucky to know him.

Olin Ashley, editor and Dad

How do I begin to thank him for all he as done for me? He gave me my writing talent, though I'll never be as good a writer as he. He gave me an ability to lavish affectionate sarcasm on those I love. He taught me the value of modesty, graciousness and pride, in my work, my family and myself. I treasure these last few months, working with him, being under his guidance once again, only this time with the full appreciation and recognition of what that means.

Donna Bachman, previous culinary instructor, boss, partner

Now Donna is my mentor and friend. She was rated one of the top five Chefs in Scottsdale, Arizona during her famous Skip's Restaurant days in the early 1990's. I was fortunate enough to be one of her students at Scottsdale Culinary Institute. I later worked for her in her fourth restaurant and eventually we dabbled in catering together. I've always been in awe of her culinary knowledge and her willingness to share her wisdom with others. If I ever have a question about a cooking technique or ingredient, I call Donna. She always knows the answer and generously shares her experience. She was a tremendous resource for me.

Susan Prieskorn, previous culinary school baking instructor

Tulsa, Oklahoma stole Susan away to lead a team of new product developers for a large commercial bakery. Our loss is definitely their gain. Fortunately, we have phones and E-mail, both getting quite a bit of exercise during my recipe testing. Susan's knowledge of baking is astounding. Her timely advice and thorough explanations saved me countless hours and numerous headaches. I know she loves her job in Tulsa, but I'm hoping she will begin to miss the desert and her friends as much as we miss her and return to the valley soon.

THE TESTERS:

Jill Rizzuto

Jill and I graduated from culinary school together and have remained close friends. She is perfecting her pastry skills at the famous Arizona Biltmore Resort in Phoenix. I'm sure she'll be running the place soon. In addition to offering continuous encouragement, she helped test several dessert recipes and a few savory recipes as well. We always have fun together and I loved working beside her once again. I respect her opinion and her talents in the kitchen (and she cleaned up after herself, too!)

Kathy Carter, Scottsdale Culinary Institute student

How does one so young acquire a palate so experienced? In her 20's, Kathy is as taste-savvy as some culinary experts in their 50's. It's amazing to me that she can decompose a bite of food into as many flavor sensations as are present in the food. Not many people can do that. I call her a super-taster. She absolutely adores food. She works 2 jobs and goes to school full-time. I enjoyed working with her, and was constantly rejuvenated by her enthusiasm and her love of our industry. She would like to be a food writer someday and I, for one, look forward to her tantalizing articles and persuasive reviews.

Daniel Domer, Scottsdale Culinary Institute student

Daniel attacked our testing with gusto and zeal. He was especially helpful to me with the wild game recipes, not only because he trimmed and cleaned the meats (one of my least favorite kitchen tasks) but more importantly, he had some experience in cooking and tasting the various species. I relied heavily on his judgment, and at his encouragement even tried to approach the tasting with an open mind. I have a much greater appreciation for wild game now than before. Daniel wants to own his own restaurant someday. I will be a regular customer and I'm sure I'll find great American cuisine on his menu, and probably a few wild game dishes, too.

THE TASTERS:

There were more than 30 tasters for the recipes in this book. Only a few have culinary backgrounds, the majority are people who just enjoy food. I'm so grateful to these friends, neighbors and associates who willingly took samples and tasted them with a seriousness Craig Claiborne would admire. Their feedback on the taste and presentation of these recipes was critical for me to provide you with a rounded view.

Rosalie and Alex Passavoy

Rosalie and Alex are very special to me. And they will be our friends for life. When I met Alex, I knew I had met a kindred soul. Alex loves to cook. He and I have made pasta together, shared seasoning secrets and commiserated over his lovely wife's unfortunate lack of taste for the crown jewels of a culinary artist, like butter and garlic and pretty much anything with flavor. I think the tasting broadened her taste buds and she actually shocked us by loving some of the recipes that explode with flavor.

Rosalie was not only helpful in the recipe tasting, but also in another area more in line with her bountiful talents. Rosalie took all the ranch logos and waved her magic wand to transform them from paper to computer, from fuzzy to sharp, from big to small and several steps in between. She saved me an inordinate amount of time during a critical stage of the book and I am grateful for her help and thrilled with the quality she produced.

Marilyn and Lou Robertson

Marilyn is an extremely competent cook. She also entertains a good bit, both family and friends. Her comments and insights were extremely valuable. She helped me disperse a ton of food as well, always funneling the feedback to me. She also was there every time I needed an extra egg or butter or spice. I think I kept her rosemary bushes clipped quite close for a time. Her husband Lou wore a path from my house to theirs, picking up food, returning dishes, and gauging my sanity.

And Others...

I also want to introduce your other tasters and thank them for their invaluable assistance. Pat and Jan Johnston, Bonnie and Art Cikens, Carol Anderson and her family, Kim and Steve Boerner, Shirley and Ervin Daskow, Malen and Pete Eyerly, Claudine and Gary Ernest, Rose Marie Maher and Carole Perlman from The Carefree Traveler, Mavis and Murry Trask, Susan and Steve Lefkowitz, and Pam Raby.

THE EXPERTS:

Betsy Hillis is an extraordinarily talented artist. She was extremely swamped when I approached her about sketching for this book. She's also a very generous and giving person, so she agreed to help me. I'm so fortunate to have her original works in this book. Now, if I could just convince her to experiment beyond white zinfandel.

Susan Meek, graphic artist and photographer, took my cover design ideas and turned them into a work of art. I appreciate her artistic ability and her eye for detail.

Sheryn Jones and her Cookbook Resources staff provided invaluable direction and guidance and increased my own publishing knowledge ten-fold. Despite the physical distance, they folded me into their virtual network and were always within a digital link.

Ron Blankenship and Bob Stachel of A.J.'s Fine Foods were tremendously helpful in providing or locating every possible ingredient dictated by the recipes. We are extremely fortunate to have such an oasis in our desert. If you are not blessed with such a culinary resource, you don't have to move to Arizona to take advantage of our treasure. A.J.'s, while not really in the direct selling business, will provide you with an alternative, and sometimes only, source of unusual or unique food products. See the "Sources" section on page 236 to see how to get in touch with Ron and Bob.

THE FAMILY:

I mentioned my husband and my Dad, but there is a "wild herd" of others who have helped me actualize my dreams and goals through their love, support and sometimes-unsolicited advice! My Mom is a remarkable woman. She can still whip out a delicious southern feast in speeds that would floor the Warner Bros. Roadrunner (and produce a mess like the Tasmanian Devil). She's always been loving, supportive and encouraging, no matter what venture I pursued. She never pushed me in the direction of the kitchen but was thrilled when I came to my senses 35 years later and pursued a culinary career.

My siblings, Jeff, Steve, Kim, Victor and Nick always provide love and encouragement and the just the right dose of humility so that I don't get too big for my britches.

I wish you could meet the extraordinary women in my husband's family, especially his grandmother Clara Walters and his aunt Sally Sliney, who first inspired me nine years ago to try stepping into the kitchen for more than just a glass of water. Their southern charm and hospitality is peerless. I can only aspire to cook and entertain in the tradition of these two culinary greats.

Last, but certainly not least, I have to thank the loves of my life, Georgia and Taylor. They kept a close vigil over the floor during the testing, sweeping in to scoop up any dropped morsels keeping my floor on the top 10 clean surfaces list. And they kept my feet warm hour after hour as I sat in front of the computer pouring my heart and soul into this book. Dogs are one of God's greatest gifts, and I'm so blessed to be owned by these two blond darlings.

Blueberry Syrup, 38
Braised Asparagus in a Lemongrass
    Broth, 114
Braised Lamb Shanks, 53

BREADS,
See also Breakfast: Breads
    Anadama Bread, 60
    Black Pepper Potato Bread, 130
    Chef Ahier's Healthy Pizza Dough, 201
    Flour Tortillas, 119
    Garlic Focaccia, 233
    Pastry Crust, 178
    Pizza Dough, 17
    Pocket Surprise Coffeecake, 82
    Swiss Cheese Popovers, 226

BREAKFAST:
  Breads,
    Banana Chocolate Chip Muffins, 176
    Blueberry Streusel Muffins, 121
    Chef Ahier's Healthy Pizza Dough, 201
    Cinnamon Raisin Bread with Maple
        Glaze, 127
    Flour Tortillas, 119
    Joyous Bran Muffins, 162
    Lemon Nut Bread, 91
    Maryanne's Sour Cream Coffeecake,
        154
    Pocket Surprise Coffeecake, 82
    Pumpkin Sweet Rolls, 105
    Sour Cream Coffeecake, 224

  Eggs,
    Breakfast Burrito in a Homemade Flour
        Tortilla, 118
    Florentine Breakfast Pizza, 202
    Herb and Feta Cheese Frittata, 191
    Smoked Salmon and Onion Omelet,
        96

  Sides,
    Almost Orange Julius, 217
    Baked Bacon, 217
    Bear Creek Granola, 161
    Broiled Texas Ruby Red Grapefruit
        with Citrus Honey and Toasted
        Coconut, 200
    Fresh Fruit Cup with Mesquite Honey,
        39
    Maple, Fruit and Nut Sauce, 147
    Oranges in Cinnamon Syrup, 145
    Spicy Apple Cider Sauce, 90
    Triple Creek Granola, 190
    Yogurt Fruit Cup, 64

Breakfast Burrito in a Homemade Flour
    Tortilla, 118
Broiled Texas Ruby Red Grapefruit with
    Citrus Honey and Toasted Coconut,
    200
Butternut Squash and Chipolte Pepper
    Soup, 183

C

C Lazy U Ranch, 71
Cajun Chicken Breast, 31
California Nori Roll, 155

CARAMEL,
    Ginger Caramel Ice Cream, 123
    "Wild Turkey" Caramel Sauce, 86

Caramelized Onion Tarts with
    Gorgonzola Sauce, 211

CARROTS,
    Lemon Carrots, 90
    Orange-Ginger Carrots, 227

CHEESE,
    Blackened Buffalo Quesadillas, 98
    Breakfast Burrito in a Homemade Flour
        Tortilla, 118

The Great Ranch
COOKBOOK

Guest Ranch Link
P.O. Box 5165
Carefree, AZ 85377
(602) 488-2202

http://www.guestranchlink.com

Please send:

_____ copies of "The Great Ranch Cookbook"   @ $19.95 each _____

Shipping and Handling   @   3.00 each _____

Subtotal _____

Arizona residents add 6.9% sales tax _____

TOTAL _____

Enclosed is a check, made payable to: Guest Ranch Link

SHIPPING INFORMATION:

Name _____

Address _____

City _____

State _____ Zip Code _____

Phone Number _____